THE
PHYSIOLOGY OF REPRODUCTION
IN THE COW

THE
PHYSIOLOGY OF REPRODUCTION
IN THE COW

BY

JOHN HAMMOND, M.A.

INSTITUTE OF ANIMAL NUTRITION
SCHOOL OF AGRICULTURE
CAMBRIDGE

CAMBRIDGE
AT THE UNIVERSITY PRESS
MCMXXVII

CAMBRIDGE
UNIVERSITY PRESS

University Printing House, Cambridge CB2 8BS, United Kingdom

Cambridge University Press is part of the University of Cambridge.

It furthers the University's mission by disseminating knowledge in the pursuit of
education, learning and research at the highest international levels of excellence.

www.cambridge.org
Information on this title: www.cambridge.org/9781107455924

© Cambridge University Press 1927

First published 1927
First paperback edition 2014

A catalogue record for this publication is available from the British Library

ISBN 978-1-107-45592-4 Paperback

PREFACE

WHEN investigating problems in the relatively new subject of Agricultural Science one is often up against the fact that there are few textbooks giving in any detail an account of knowledge up to date, so that while writing an account of an investigation into the physiology of reproduction and the development of the udder in the cow it was determined to include references to recent literature on the subject. In this way it was hoped that the account would be of use in the future to those doing research on this subject and save them time in getting acquainted with the literature as well as forming a reference book for advanced students and practitioners in Agricultural and Veterinary Science.

Unfortunately there was some little delay in getting a work of this kind with many illustrations published and since the text was written in 1922 many papers have been seen which should have been referred to; as it was difficult to insert these in the text mention has been made of the more important ones in an Addendum.

My thanks are due to Dr F. H. A. Marshall, F.R.S., who has throughout the course of the investigation given me the benefit of his advice and wide experience.

I am also under obligation to the late Mr K. J. J. Mackenzie, Director of the University Farm, who put his records and many of his animals at my disposal.

My thanks are also due to Messrs Warrington and Sons for the use of their slaughterhouse at all hours of the day and night, and to Mr N. Langridge of the University Plant Breeding Farm, for assistance in the selection of suitable animals.

I am indebted also to the Royal Veterinary College for the use of their Library; to the willing help of my laboratory assistants—S. J. Tadman and J. Pike, on whom many of the arduous duties have fallen, and to Clement W. Williamson who has taken the photographs for me.

The experiments were conducted at the Field Laboratories, Milton Road, Cambridge.

The expenses of the investigation were largely defrayed by a grant from the Ministry of Agriculture and Fisheries who have also provided funds to assist in publication.

J. H.

CONTENTS

LIST OF ILLUSTRATIONS

PLATE VII

PLATE VIII

PLATE IX

PLATE X

PLATE XI

PLATE XXIV

1. Section through the foetal membranes on the outer edge of a cotyledon from the uterus of heifer P 9 in the 8th month of pregnancy. For description see p. 151. Scale, $\frac{1}{6}''$ and 3 ($\times$ 8) eye-piece.

2. Section through the foetal and maternal projections of the cotyledon on the inner (uterine) side; from heifer P 9 in the 8th month of pregnancy. For description see p. 151. Scale, $\frac{1}{6}''$ and 3 ($\times$ 8) eye-piece.

3. Section through a portion of the intercotyledonary mucosa from the uterus of heifer P 4 in the 2nd month of pregnancy; showing the erosion of the uterine epithelium where the foetal membranes press against its surface. For description see p. 156. Scale, $\frac{2}{3}''$ and 3 ($\times$ 8) objective.

4. Section through an accessory cotyledon from the uterus of heifer P 7 in the 6th month of pregnancy. For description see p. 156. Scale, 2$''$ and 3 ($\times$ 8) eye-piece.

PLATE XXV

1. Photograph of the foetus and foetal membranes taken from the uterus of heifer P 2; pregnant 1 month. For description see p. 158. Scale, $\frac{1}{2}''$ marks at base of photo.

2. Photograph of the foetus and part of the foetal membranes taken from the uterus of heifer P 4; pregnant 2 months. For description see p. 159. Scale, $\frac{1}{2}''$ marks on right side of photo.

3. Photograph of part of the foetal membranes taken from the uterus of heifer P 3 in the 3rd month of pregnancy, showing the part lying in the non-pregnant horn, body and a portion of that in the pregnant horn. The umbilical cord has been cut and the amnion laid out to show the developing amniotic pustules. For description see p. 159. Scale, $\frac{1}{2}''$ marks on right side of photo.   .   .   .   *facing page* 158

PLATE XXVI

1. Section through the cervical epithelium of heifer P 9 in the 8th month of pregnancy. For description see p. 165.

2. Section through the wall of a follicle in which blood had penetrated into the liquor folliculi; from the ovary of cow A 4.

For description see p. 184 Scale, $\frac{1}{6}''$ and 3 ($\times$ 8) eye-piece   *facing page* 164

PLATE XXVII

1. Transverse section through a cystic Gärtner's canal in the vagina of cow No. 28 which had cystic ovaries. For description see p. 98. Scale, 2$''$ and 3 ($\times$ 8) eye-piece.

2. Section through one of Gärtner's canals in the upper part of the vagina of heifer P 3 in the 3rd month of pregnancy. For description see p. 168. Scale, $\frac{2}{3}''$ and 2 ($\times$ 6) eye-piece.

3. Transverse section through one of Gärtner's canals in vagina of heifer C 4; 48 hours after beginning of heat. For description see p. 98. Scale, $\frac{2}{3}''$ and 1 ($\times$ 5) eye-piece.

4. Section through a fold in the epithelium of the vagina next the os; from heifer P 9 in the 8th month of pregnancy. For description see p. 166. Scale, $\frac{2}{3}''$ and 3 ($\times$ 8) eye-piece.

5. Section through a lymphatic nodule in the vaginal mucosa of heifer P 2 in the first month of pregnancy. For description see p. 166. Scale, $\frac{2}{3}''$ and 3 ($\times$ 8) eye-piece.

PLATE XXVIII

1. Vertical longitudinal section through one side of the udder of heifer P 3 killed at the 3rd month of pregnancy. For description see p. 172. Scale, $\frac{1}{2}''$ marks at top of photo.

I

INTRODUCTION

THE study of the reproductive processes in the cow was undertaken with several practical objectives in view.

Veterinarians have for some time past been treating cows for sterility by various empirical methods, some of which have been quite successful at times, but there has been very little work done on the physiology of reproduction in the cow to form a basis for the rational treatment of such cases.

A study of the changes occurring during pregnancy was attempted in the hope that it would serve as a basis for the study of the pathology of the organs in abortion, and that it might also indicate the causes which led to the development of the udder and those which may form the foundation of the genetic differences between good and poor milking cows.

Although artificial insemination has been found useful in mares few attempts have been successful in cows, and it was thought that a detailed study of the reproductive processes in the cow might show the cause of failure and lead to extensions of this method of reproduction.

The experiments were on the whole designed to elucidate problems which if solved would do much to assist the farmer in regulating the management and feeding of his herd for the production of calves and milk.

From a scientific point of view it was thought that a comparison of the processes occurring in the reproductive organs during pregnancy with those occurring during the cycle in the same species would throw light on the processes involved in the cycle, and enable one to realise the significance of these changes.

The similarity between the reproductive cycles in the cow and in woman (the former having a cycle of three weeks and the latter of four weeks), which in both cases culminate in the flow of blood from the vulva, renders the study of reproduction in the cow particularly interesting to those engaged in elucidating problems in human reproduction and the relation of "heat" in animals to menstruation in women.

One of the chief difficulties met with, and probably the reason why such a study has not been undertaken before, is that the cow is an expensive animal to keep and kill, so that the numbers used for experimental purposes must necessarily be limited. For this reason the experiments made have been precise detailed studies of a few individuals rather than a superficial examination of a large number. The size of the reproductive organs is of great advantage in the study of reproduction, for observations with the naked eye can be made on the ovaries which are not possible in the fresh condition in the mouse or rat, where resort has to be made to staining and microscopical examination. Measurements and weights can also be made and taken to a degree of accuracy which is impossible in a small animal. Moreover, since as a rule only one egg is shed at each ovulation, the process of tracing the corpus luteum during its history, and of experimenting with it, was much easier than if several were shed as in the dog, ferret and rabbit.

Discussions of the sexual life of the cow and other domestic animals have been published by Schmaltz [1] and Kronacher [2] in whose treatises many of the problems investigated have been outlined.

MATERIAL.

The material on which this study is based falls under two headings: (1) experimental animals, (2) organs collected from slaughterhouses.

(1) The majority of experimental animals used were cross-bred Shorthorn Heifers of 2–3 years old. Heifers were selected rather than cows so as to eliminate chances of infection of the uterus and organs by contagious abortion or other diseases occasioned by pregnancy. The facilities for keeping the animals were limited and lots of about four only could be accommodated at any one time; this had the advantage, however, that it was possible to keep them all under close observation. The majority came from the Cambridge University Plant Breeding Farm (and my thanks are due to Mr N. Langridge for the assistance he has given me in the selection of these animals), where they had been kept a year or more, and a few were purchased from a dealer direct. All were kept under observation for a month or two before the experiments proper began.

The few cows that were used were not purchased on the market but came direct from farmers in the district from whom an account of their previous history was obtained.

(2) Material was also collected from the slaughterhouses of local butchers (and I have to thank more especially Messrs Warrington and Sons for their assistance in this respect). In the majority of cases, however, the previous history of these animals was not known as most were pur-

chased in the open market; in some cases, however, where the animals came from the University Farm or from local farmers the breeding history of the cow was obtained.

In addition to the foregoing material much information was obtained from experienced stockmen met at the Agricultural Shows (see Appendix I). Many have had a life-long experience in the management of breeding cows and to them my thanks are due for various details which have suggested new lines of research.

METHODS.

An attempt has been made to study the oestrous cycle by timing the processes concerned by means of time observations, in much the same way that physiologists have studied muscle-nerve preparations by the use of recording drums.

Owing to its importance from a practical point of view it was necessary to know exactly the duration of oestrus. While by observation it was generally possible to tell whether a cow was on heat or not, yet it was thought best to apply the test of whether she would actually "take the bull" or not. In order to do this without interrupting the cycle by a pregnancy ensuing, a vasectomised bull was used; such an animal has all the desire and powers of coitus possessed by an entire animal but ejects no spermatozoa.

The bull was kept in a loose box separate from the cows, but, except for a few experiments, within sight of them. To determine whether or no a cow was on heat the bull was turned into an enclosure or covered shed with her and they were kept under observation. It was decided to test the cows at intervals of every two hours, commencing a day or more before the expected heat period was due, and if there was any doubt (as sometimes happened when they were coming on and going off) a period of half-an-hour was allowed before the bull was withdrawn. In this way it was considered that the actual time a cow came on or went off heat could be determined within about an hour. Extracts from the log-book kept are given in Appendix II.

After the experiments had been going on for some time and it was found that the cow was staying on heat 12 hours or more, the bull was not put in at the 4th, 6th or 8th hour after heat began, but was reintroduced at the 10th and subsequent 2 hour periods; in this way his energies were conserved, and the end period of the heat could more readily be determined.

This method has involved continuous observation of the cows over considerable periods, and my thanks are due to Mr S. J. Tadman, my laboratory assistant, who has shared this work with me, one of us watching by day and the other by night.

This work has been supplemented by a study of the anatomical and histological changes in the reproductive organs of these animals, which were killed at different periods of the cycle and of pregnancy. An accurate reproductive history of the animals was, therefore, obtained before they were killed. The observations were made both in the fresh state and after they had been fixed (for the most part in 10 per cent. formalin solution) and stained (mostly in iron or Delafield's haematoxylin and eosin).

The bull used was operated on when he was about 15 months old, having previously been used to get four heifers in calf. He was killed when about four years old, after the experiments were completed; during this time he remained sexually active, mating on the average about five times a week. His secondary sexual characters were normal, and the testes as far as could be judged were of normal size. The operation was performed by making two small incisions (at the top of the posterior side of the scrotum) through the skin, cremaster muscle and tunica vaginalis, which exposed the vas deferens; a small portion of each of these latter was then removed but the ends were not ligatured and the wound was stitched up.

The testes were sectioned after the animal was killed and were to all appearances normal, both seminiferous tubes and interstitial cells being well developed and the former containing spermatozoa. This result does not agree with Ancel and Bouin's[3] experiments with vasectomised horses and other animals, in which they found degeneration of the seminiferous tubules and hypertrophy of the interstitial cells. They ligatured the vas deferens, however, and thus by causing pressure and inflammation in the tubules may have aided their degeneration in much the same way as the mammary gland atrophies when the products of secretion accumulate within it.

Steinach[4] states that the interstitial cells of the testis undergo hypertrophy after vasectomy and that the seminiferous tissue disappears.

Lipschütz[5] however believes that no actual hypertrophy of the interstitial cells occurs and that whether or no degeneration of the seminiferous portion is found after operation depends on the interval between the operation and the date of killing.

II

THE BREEDING SEASON

Data collected from Milk Recording Societies soon established the fact, as is generally known, that cows will breed at all times of the year. These data which are given in Table I show the percentage of the total number of cows which calve in each month of the year; it will be seen that the time of year that they actually calve depends on local agricultural conditions and is no guide to their natural breeding season, for in the west of England, where grassland prevails, the majority calve in March and April, whereas in the east on arable land, where milk is required for winter supply to the towns, more calve in the autumn, although even here a large number tend to calve in the spring.

Weber[6] found that in domestic cows heat occurred at all times of the year.

Heape[7] states that whereas domesticated cattle breed all the year, in wild cattle the calving season is limited, but in the Zoological Gardens they are capable of breeding at any time of the year.

Mayo-Smith[8] found that in man the largest number of conceptions occurred in the summer months, and the time of maximum number varied with the climate, in Greece the maximum occurring in April, whereas in Sweden the maximum did not occur until June.

Statistics collected from various authorities are appended to this table (I) and they confirm the above conclusions; in particular there are large differences between Ayrshire and Denmark in the time of year cows usually calve.

In order to determine the readiness with which cows would breed at different times of the year the average length of time was calculated between the times cows calved and their next fertile service. Data on this point are shown in Table II. The average time is 97 days but cows which calve in May are, on the average, served within 76 days, whereas those which calve in September are not, on the average, served until 135 days after calving, and there is a gradual and steady rise and fall between these two extremes. In order to calve on the same date in the following year a cow should be served 85 days after calving. The average interval before service throughout the year varied in different districts, being as high as 105 days in Penrith, and as low as 89 days in Norfolk.

(*a*) *The time of year.* From Table IV it will be seen that the average length of the heat period is greatest in the warm summer months and shortest during the cold winter months, there being an average difference of 5–6 hours between the two extremes. Herdsmen (Appendix I (2 *b*)) have noticed this difference and some say that a cow can be made to go off heat by throwing cold water over it.

Detailed observations of the signs of heat, such as the flow of mucus and activity in jumping others (see p. 29), showed them to be much more marked during the summer than in the winter months. The observations given in Table IV are not affected by this last circumstance, as the bull was used to test the animals and frequently in winter it was found that he would serve a cow when no very obvious signs of heat were present. The short duration and slight signs of heat which occur in the cold months of the year may account for difficulties frequently met with by farmers in getting cows served during the winter months, more especially where the bull is not run with the cows. A heat period of 6 hours occurring either during the day or night is easily missed and one of 10 hours would be frequently overlooked as in winter the men's working hours are short.

Heape (7) states that in the bitch the winter oestrus in some breeds does not last so long as the summer oestrus.

(*b*) *Age.* Table IV shows that cows (19·3 hours) have on the average a slightly longer duration of heat than heifers (16·1 hours). In Table III (*b*), however, where the differences in age are slight, any change in this direction is masked by individual differences. No opportunity has occurred of testing the same animal at different ages. The experience of herdsmen (Appendix I (2 *b*)) however is that there is no marked difference in the duration of heat between cows and heifers.

(*c*) *State of fatness.* The evidence given in Table III (*c*) is in favour of the duration of heat being shortened by an animal being in a fat condition, this change being observed both in cows and in heifers. This fact may account for some of the difficulty met with in practice of getting fat animals to breed, since if the oestrus is shortened there is more chance of it occurring during the night only and so being missed. Weber (6) found that fattening foods given for a short time produced no effect on the heat period but that if long continued the intensity of heat was diminished. Marshall and Peel (53) found a considerable quantity of lipochrome in the ovaries of fat heifers. Possibly the decreased length of the oestrus in very fat animals is caused by a blockage of the ovarian cells with lipoids (see p. 82).

(*d*) *Individuality*. Individual differences are very marked as will be seen by Table IV, the heifers varying from 8 to 21 hours and the cows from 17 to 21 hours. With the latter the range is probably larger as only a few individuals were experimented with. Individual differences in the length of the oestrus is probably one of the causes of the "shy breeders" that are frequently met with.

(*e*) *Effect of drugs*. If drugs with an effect of prolonging the heat period could be found they would prove very useful. The various aphrodisiacs on the market are said to stimulate heat but there is very little definite knowledge of their action either in accelerating the onset of the heat period or of prolonging its duration.

Yohimbine given for 5 days (Cow A 3, Table IV) in doses of ·5 gm. per day and then for 8 days in doses of 1 gm. per day and continued during the time the cow was on heat failed to influence the duration of heat; she was "on" 20 hours when treated with the drug as against 20 and 22 hours in the previous and succeeding periods. This cow came on heat at 8 a.m. on April 1st and was dosed with ·5 gm. yohimbine at 9 a.m. and 6 p.m. on March 31st and at 9 a.m. and 6 p.m. and 10 p.m. on April 1st, but went off heat at 4 a.m. on April 2nd.

(*f*) *The psychological effect of proximity of the male* and other animals had no effect in prolonging the heat period as will be seen from Table III (*d*). The length of the heat period was not altered by keeping the heifers apart from the bull (C 2 and C 3) nor was it affected by keeping the animal apart from other heifers (C 7). The signs of heat are however affected by this factor (see p. 29).

(*g*) *The effect of service by the bull*. In some cases heifers were served every two hours during the heat period and in others only once as soon as they came on and then left for a number of hours to see if they would go off more quickly. Heifers were also tried when the bull was not allowed to serve them at all during the oestrus. In the latter experiments in two cases the heifers went off heat quicker after service than if not served at all, but in the third case this was not so. Some herdsmen (Appendix I (2 *b*)) believe that cows stop on heat longer if not served. Weber[6] however could find no difference in this respect. Robinson[51] found in ferrets (which do not ovulate spontaneously) that fertile coitus caused ovulation and so reduced the time the animal remained on heat; this is also the case in rabbits[24]. Although the cow ovulates spontaneously it may be that in this animal ovulation is accelerated by coitus; further work however is required before this can be taken as proved. Marshall[54] found in the sheep that, although ovulation occurs spontaneously at each of the earlier heat periods of the sexual season, coitus is necessary to cause ovulation towards the end. He also found that coitus hastened the rupture of the follicle.

(*h*) *The side on which ovulation occurs.* Only one case was observed in which ovulation occurred on the same side as that in which the previous corpus luteum had formed and Table III (*f*) shows that this did not appreciably affect the duration of oestrus.

(*i*) It has been stated that the act of milking shortens the duration of heat and that cows are best put to the bull before they are milked; the opinion of the majority of herdsmen however is against this view (see Appendix I (5)). No experiments have been made on this point, however, and it may be that the belief has arisen owing to the fact that in the morning the cows have often not been seen for 16 hours and would frequently be "off heat" if left until after milking.

(*j*) *Experimental removal of the corpus luteum.* One of the most striking results of the removal of the corpus luteum is in the duration of the subsequent heat period. In both experiments which have been done the corpus luteum was squeezed out about 7 days after the last heat period, and the next heat period occurred about two days after the operation or 9 days after the last oestrus; thus the normal length of the cycle was halved (see above, p. 15). In both these cases the duration of oestrus was also halved, cow A 3 being on heat for 10 hours instead of the average 21 hours, and cow A 5 being on heat for 12 hours instead of the normal 19 hours.

This result is one of fundamental importance as regards the cause of oestrus. Quite how the facts should be interpreted is in question and will be discussed below.

That the duration of the heat period is correlated with the length of the preceding cycle is a point which has hitherto not been considered. As far as it relates to different species of animals there can be no doubt that no correlation exists as the following table shows:

Species	Length of cycle (days)			Duration of heat (days)		
	Marshall (55)	Curot (22)	Franck-Albrecht (56)	Marshall (55)	Curot (22)	Franck-Albrecht (56)
Mare	21	28	21–28	4–5	7–10	8
Cow	21	21	21–28	1	$\frac{1}{2}$–1	1–2
Ewe	15	21	21–28	Few hours– 1 day	1–3	1–1$\frac{1}{2}$
Sow	21	21	14–18	1	4–8	1 day to 40 hrs.

Yet the experimental results given above seem to show that within a species the length of the cycle and the duration of the subsequent oestrus are correlated.

Marshall(54) found that in sheep when the number of oestrus periods is increased (by bringing highland sheep to the lowlands), the duration of oestrus is shorter and that in Dorset Horns where the oestrus may occur every 11 days (instead of the usual 13–18) the oestrus may only last about 2 hours.

Table V gives the length of the cycle compared with the duration of

Table V. *Correlation between the length of cycle (hours) and the duration of the subsequent oestrus (hours).*

Group of cycle lengths hours	Individual cycle lengths	Associated duration of subsequent oestrus	No. of observations	Average cycle length	Average oestrus length
Over 480	576, 566, 544 538, 538, 516, 510 508, 502, 502, 506 492, 490, 480	20, 20, 10 22, 20, 24, 14 22, 20, 18, 18, 10 18, 22, 18	15	518	18·4
460–479	474, 470 468, 464 462, 462	18, 12 18, 20 18, 26	6	467	18·7
440–459	458, 458, 458 456, 456, 455, 454 452, 448 442, 442, 440	16, 16, 16 18, 8, 14, 22 16, 14 14, 30, 10	12	452	16·2
420–439	438, 436, 436, 436 432, 432, 430, 428 426, 424, 424, 424 422, 422, 422, 420 420	18, 12, 16, 22 12, 20, 20, 16 6, 8, 14, 22 10, 6, 22, 12 14	17	403	14·7
Under 420	412 212, 196	16 12, 10	3	307	12·7

the subsequent oestrus in all the observed cases (53) in heifers and cows; it shows that a sensible correlation exists. The cases "over 480 hours cycle" do not show any increase over the "460–480 hours cycle," but the number of observations made in the latter group is not very large. It may be, however, that in some cases the length of the cycle is frequently prolonged not by the persistence of the corpus luteum but by conditions being unfavourable for the maturation of the follicle after the corpus luteum had atrophied. Hess(41) states that in cows after infection with granular vaginitis the corpus luteum often hypertrophies, and that in these cases the heat periods are often abnormally long.

From our results it would appear that the length of oestrus (desire) depends on the degree of hypertrophy of the uterus and other generative organs brought about by the action of the previous corpus luteum, and that if this is cut short by removal of the corpus luteum at an early age the length of the subsequent oestrus is reduced.

There is, however, much to be said against this view. Robinson[51] found with ferrets that the length of oestrus depended on the time ripe follicles remained in the ovary, and that coitus shortened the length of oestrus by causing rupture of the follicles. This might be explained in our opinion by the formation of the new corpora lutea (as a result of follicular rupture) inhibiting the condition of oestrus. Loeb[57] found that in guinea-pigs the presence of the corpus luteum may prevent the signs of heat although apparently mature follicles are present in the ovary. In Marshall and Runciman's[58] experiments with bitches, artificial rupture of the follicles before heat did not prevent the onset of oestrus, but in this case normal corpora lutea were not found in the ovaries. In the cow the actual bursting of the follicle can have little effect on the length of oestrus (desire) for ovulation does not normally occur (see p. 34) until 24–48 hours after the commencement of desire, whereas oestrus itself lasts only about 17 hours. The two cows A 3 and A 5 afforded an illustration of this.

In the case of A 3 the corpus luteum was squeezed out 6 days after the last heat, and the next oestrus occurred 53 hours after the operation and lasted for 10 hours only, although the follicle which ripened at this time had not ruptured when the animal was killed 6 days later. In the other cow, A 5, the corpus luteum was squeezed out 7 days after the last heat, and the next oestrus occurred 48 hours after the operation and lasted for 12 hours; the follicle in this case had ruptured when the animal was killed 13 days later.

Hess[41] found that 95 per cent. of cows if served become pregnant at the first heat after squeezing out the corpora lutea.

These two cases show that the duration of heat is much the same whether the follicle ruptures or not. Strodthoff[59] from the results of rectal examination of ovaries in the cow concluded that ovulation does not always occur with heat.

Another interpretation of the results obtained from these two cows (A 3 and A 5) is nevertheless possible. It is that although heat occurred in each at about the same time after removal of the corpus luteum and lasted for about the same time, yet the ripening of the follicle and ovulation did not occur until some time afterwards, and the cow A 3 was killed before it had occurred. In other words, by removal of the corpus luteum the external symptoms of heat were obtained, but the ripening of the follicle and ovulation did not closely follow these as it does under normal conditions (compare with cycle in woman, see below).

In the experiment on cow A 3 there is doubt as to whether a cyst was formed as a result of squeezing out the corpus luteum or whether although heat occurred ovulation was delayed several days.

The evidence however, as far as it goes, is rather against the latter view, for the unruptured follicle in A 3 was above normal in size (but this may have been due to pre-ovulation swelling), and the appearance of the corpus luteum in A 5 indicated a probable ovulation soon after the heat was observed (see further under Sterility, p. 190).

The possibility remains that heat (desire) is determined by some stage in the maturation of the follicle and that whether it ruptures or not is immaterial. Robinson(51) found that in ferrets oestrus only appears when the follicles attain a pre-inseminal stage, and that before they rupture an additional swelling of the follicle occurs. It is shown below (p. 41) that in the cow the follicle does swell after "heat" is over and just before it bursts (but not to the extent found by Robinson in ferrets) and it is conceivable that the swelling might terminate the heat. A possible explanation of our results might therefore be that the "heat stage" of the follicle is more quickly passed through if ripening is fast (as occurs when the corpus luteum is squeezed out), but this explanation would not fit in with Marshall and Runciman's(58) experiments in bitches, in which heat occurred at the normal time after the follicles had been artificially ruptured, nor is it compatible with the phenomena of oestrus and ovulation in bats, in which oestrus occurs in the autumn but ovulation not until the following spring.

"Desire" is probably, physiologically speaking, a nervous state induced reflexly, and in man owing to high nervous organisation it may not require the intense internal stimuli for the reproductive organs as is necessary in the case of the lower animals. There is not much evidence at present as to whether the desire is caused directly by the action of the primary factor in the ovary (follicle, interstitial gland), or whether it is occasioned by secondary changes (due to regression of corpus luteum or stage of follicular development) set up in the accessory glands and erectile organs of the vagina and vulva by the primary organs. It is not due to any uterine changes for it has been shown that hysterectomised rabbits copulate normally(60). If in the above experiments (squeezing out corpora lutea) the primary effect of the removal of the corpus luteum in causing the heat is objected to on the grounds of experiments performed by other investigators, it might be reasonable to suggest that the duration of heat may be caused by its secondary action on the upbuilding of the accessory glands and erectile organs of the vagina and vulva. Unfortunately the cows in which the corpora lutea were squeezed out (A 3 and A 5) did not normally bleed after their oestrous period (as frequently occurs in heifers, see p. 57) so that it is not known whether the "desire for coitus" at heat is necessarily associated with the phenomena of swelling and congestion of the reproductive organs.

Fraenkel (42) found that destruction of the corpus luteum in women prevented the appearance of the next menstruation at the normal time and he therefore attributed to the corpus luteum the function of causing the flow. Halban and Köhler (61) have also performed this operation and have criticised his results; they found that removal of the corpus luteum is followed by bleeding (rather less than the normal menstrual flow) 2 3 days after the operation and then the regular occurrence of menstruation at the usual interval (28 days) after the post-operation bleeding, the menstrual rhythm being upset by the removal of the corpus luteum. They also found that the duration of the post-operation bleeding did not depend on the age of the corpus luteum which was removed, and concluded that menstruation was prevented by the corpus luteum up to the time of its maximum development, and that although the corpus luteum regulates the duration of the cycle, menstruation was not caused by it but by some other ovarian hormone. In three cases however they removed not only the corpus luteum but also both ovaries and these cases were followed by a menstrual flow 2–3 days after the operation, but no further menstrual periods occurred. It appears to us that the results of these three cases suggest that it is the removal of the corpus luteum which causes the bleeding rather than the presence of another ovarian secretion (see also Weymeersch (62)).

Ancel and Villemin (63) found that in women ovulation occurs 12 days before menstruation, and that the latter occurs at the time of maximum development of the corpus luteum.

Schröder (64), Miller (65), and Marcotty (66) all concluded that ovulation in women does not occur until about the 12th or 14th day after menstruation and that the onset of the latter generally coincided with the regressive phase of the corpus luteum. Leopold and Ravano (67), however, concluded that ovulation occurs most frequently at menstruation but they also found that menstruation can take place without ovulation. Siegel (68) who collected statistics of the influence of the time of coitus on fertility in women, found that the likelihood of fertilisation increases from the beginning of menstruation, reaches a high point 6 days later, remains almost at the same height until the 12th–13th day and then declines to the 22nd day after which there is absolute sterility. Ovulation, he estimates, occurs between the 11th–15th days.

For a review of recent work on this point the reader is referred to articles by Novak [69] and Triepel [70]. Heape [71] found that the ovaries of menstruating monkeys do not always contain follicles in a state of approaching ripeness.

In the bitch bleeding occurs just before ovulation, in woman about 12 days before it, and in the cow at about the time of ovulation. It may be that the degree of atrophy of the corpus luteum which is necessary for the maturation and bursting of the new follicle differs from that which is necessary to cause bleeding in the different species, and so the processes are not always correlated in point of time. In the same way the time relations of desire for coitus and bleeding do not always coincide in animals, for in the bitch "desire" occurs after bleeding, whereas in the cow it occurs before bleeding. It is difficult to see how this can occur if both bleeding and desire are dependent on the ripening of the follicle.

The general impression formed from the results of our own experiments considered in conjunction with those obtained by others in the bitch, ferret, and woman, is that the internal secretion of the follicle and of the corpus luteum is essentially the same and not different as some suppose (see p. 80). Meyer [72] concluded that both the ripe follicle and the proliferation stage of the corpus luteum stimulate the regeneration and pathological growth of the uterine mucosa. Both may give rise to uterine hypertrophy, and they only differ in the amount of the internal secretion that obtains entrance to the blood stream; the follicle in most stages of its development stores the active principle (possibly liquor folliculi) which obtains entrance to the blood only from the network of capillaries in the theca; in the corpus luteum however the luteal cells are in close connection with the capillaries so that the secretion is not normally stored but passes to the blood stream. This secretion probably acts on the uterine blood vessels causing their dilation and bringing about the hypertrophy of the uterine mucosa in preparation for pregnancy; the sudden removal of the secretion causes collapse of the blood vessels and extravasation of blood with disintegration of the mucosa.

In the monoestrous bitch the simplest form is seen where follicular action causes pro-oestrus or heat hypertrophy (similar hypertrophy is seen in the cow—nymphomania—when cysts are present) and a persistence of a similar condition is observed in the ferret when coitus does not occur. The formation of the corpus luteum in the bitch has been shown by Marshall and Halnan[73] to be attended with uterine hypertrophy, and they found extravasated blood in the stroma of the uterus at the end of the pseudo-pregnant period. Craig[17] has stated that at the end of the pseudo-pregnancy in bitches a swelling of the vulva with reddening of the vaginal mucous membrane occurs. During normal pregnancy uterine hypertrophy occurs under the action of the corpus luteum, and at parturition similar but more marked changes than those at the end of pseudo-pregnancy take place.

In the cow and polyoestrous animals generally the changes that occur during heat and pseudo-pregnancy are telescoped so that the regression of the corpus luteum overlaps the ripening of the next follicle. The same thing occurs in woman, but the overlapping is not so great as in the cow. In both these cases bleeding will occur after that stage of the cycle at which the greatest hypertrophy of the uterus occurs; this in the cow happens about the 20th day after the previous ovulation and about the time of the second ovulation; but in woman about the 15th day after the previous ovulation and 13 days before the second ovulation as the following sketch shows:

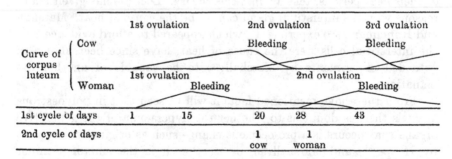

In this connection it may be noted that an intermenstrual pain sometimes accompanied by a discharge has been described in women (Croom[74]). It is possible that this is associated with the ripening of the follicle.

Thus it will be seen that difference in the length of the cycle in the cow and in woman is due not so much to differences in the length of life of the corpus luteum in the two species as it is to the degree of luteal regression necessary before a new follicle can ripen. In the rat where the cycle is very short (4 days), the diminution in size of the corpus luteum does not usually take place until the beginning of the third cycle succeeding a corpus luteum of ovulation (Long and Evans(32)), although the luteal cells are in a state of regression at the end of the oestrous cycle which gave it origin. The essential conditions necessary to cause bleeding appear to be the sudden removal of the stimulus causing hypertrophy of the uterine mucosa. That bleeding is not due to the mature follicle in woman is shown by the fact that the ripening of the follicle does not occur until after menstruation has ceased. Weymeersch(62) has shown that removal of the corpus luteum in rabbits pregnant 8–10 days caused constriction of the uterine blood vessels, oedema and extravasation of blood into the tissues of the uterine mucosa. Stockard and Papanicoloau(38) concluded that in guinea-pigs the absence of the corpus luteum causes menstruation, and that spayed females have a long and continuous atypical destruction of the uterine mucosa.

The theory that the interstitial cells are the cause of heat would not explain the alteration in rhythm of the cycle caused by squeezing out corpora lutea, nor would it account for the occurrence of a menstrual flow in women immediately after the removal of the corpora lutea from the ovaries and the subsequent regularity of their cycles dating from this flow. Moreover the presence, absence or varying quantities of interstitial tissue in different species, bearing no relation to the frequency of their heat periods, tells against this theory. Loeb(75) has given many reasons why the interstitial gland cannot be the cause of heat. Marshall and Runciman's(58) experiments, which appeared to afford evidence that the interstitial cells were the cause of heat, have since been shown in extended experiments by Marshall(76) to be capable of another explanation.

From the above mentioned facts it will be seen that in polyoestrous animals there is doubt as to whether the hyperaemia of the generative organs that occurs is a pro-oestrous change such as originally described by Heape(77) and Marshall(78), or whether it is the end of a minor pseudo-pregnancy determined by the atrophy of the corpus luteum.

In the monoestrous animals (bitch—Marshall and Jolly(79), Marshall and Halnan(73)) both occur and it would appear probable that in polyoestrous animals owing to the shortness of the cycle the two nearly coincide in point of time.

More experimental investigation is required, however, before definite conclusions can be arrived at. Further work on the effects of removal of the corpus luteum (with time observations) from heifers which bleed after oestrus would furnish valuable results not only as to the origin of heat and diagnosis of cases of sterility, but also in explaining the menstrual cycle in women.

3. Psychological Changes—External Signs of Oestrus.

Various outward and visible changes occur in an animal during the time that it is "on heat." These vary considerably in different individuals and at different times of the year but chiefly in degree rather than in kind.

No notable differences in the behaviour of cows exist during the greater part of the oestrous cycle but changes occur as soon as "heat" appears.

The following description of the signs of heat is based on close observation of more than 60 periods in about 20 different animals. For examples of individual periods see Appendix II.

The various symptoms are displayed best when the bull or other cows are present. At the onset of heat a cow becomes restless and frequently when tied in the stall is seen standing while the others are lying down. She twitches her tail frequently and often raises it. When out at grass the cow does not usually continue to feed but wanders about the field and frequently goes off by herself or with another cow which she rides or is ridden by. She will also frequently jump the bull and when he attempts to serve her will stand still and not move away as happens when the animal is not on heat. A cow on heat also frequently lowers the hips and small of the back and raises the tail head; she also, more especially after jumping other cows (and after service by the bull), arches her back and strains, a string of glairy mucus coming away from the vulva at this time.

When on heat she will also frequently play with the bull by horning him or will stand and lick him, and when separated from other cattle will low or "blar." The bull will "hang round" a cow that is just coming on or going off, although she is not actually on heat at the time, and will not stand to service.

A cow differs from the majority of other animals (sow, bitch) in that there are no marked pro-oestrous changes of the reproductive organs visible to the naked eye. Occasionally a little clear mucous fluid issues from the vulva an hour or two before heat begins but this does not always occur. No pro-oestrous congestion or swelling of the external genital organs occurs before the oestrus.

During heat a large flow of mucus occurs from the vulva and runs down the tail or flanks. The character of the mucous flow changes as heat proceeds; at first it is clear and very fluid, later it contains yellowish cheesy lumps and afterwards it becomes whitish and thickened (see p. 99).

During oestrus the vulva becomes slightly swollen and flushed but this does not usually reach its maximum development until about a day or two after the heat is over. Very often, especially in heifers it cumulates in a flow of blood about 2–3 days after the commencement of oestrus. The occurrence of bleeding and its significance will be discussed below (p. 57).

The observations made showed that as a rule the symptoms of heat are more marked in the summer than in the winter and this is confirmed by the opinion of herdsmen (Appendix I (2 *b*)). The symptoms are also more marked when the cow is running loose with others than when tied up in the stall and also when the animal is warm rather than when she is cold. Like other reflexes in the body the reaction appears more marked when the animal is warmed up.

There is usually a short period at the beginning of heat during which the animal is coming on and at the end a rather longer period when she is going off; at these times the symptoms of heat are less marked but during the remainder of the period no difference in the intensity of heat could be observed.

Some animals do not exhibit such intense signs of heat as others (probably depending on the nervous temperament of the animal), and herdsmen say (Appendix I (2 *b*)) that in certain breeds the signs of heat are very slight; they also believe that heifers are as a rule more excited at heat than old cows.

After coitus the reproductive organs appear to undergo rhythmical contractions which possibly assist the entrance of the spermatozoa into the uterus. Busquet[80] has pointed out that this also occurs in other species.

Weber[6] has studied in great detail the symptoms of heat in cows and has collected the literature on the subject. He divided his animals into groups: (1) those with intense heat symptoms, (2) those with moderate heat symptoms and (3) those with feeble heat symptoms; but while this is true on the whole the differences due to time of year and circumstances make this arbitrary division almost impossible in practice.

He states that the behaviour of cows on heat is very irregular, that only about 38 per cent. bellow, and that most are in a nervous irritable condition, many lifting and switching the tail and depressing the back. The jumping on other cows always occurred and a cow on heat would stand still while others jumped her. Tied up cows frequently showed heat by dragging on their head chains and licking other cows and their attendants. Only animals with intense heat periods went off their food at this time but he could observe no change in the frequency of micturition and defaecation in such cows.

He found that reddening of the vaginal mucous membrane, swelling of the genitals and a flow of slime from the vulva always took place at heat but that these symptoms varied with the intensity of the heat exhibited by the cow. These symptoms became less marked as the age of the cow increased. Rarely blood appeared in the secretions during heat but in 82 per cent. of the cows it appeared about 1–5 days after heat and was much more frequent in heifers than in cows. Swelling of the udder occurred only rarely and the flow of milk was affected in a few instances only[81]. One case has come under our notice where the yield of milk was considerably reduced during heat. The rectal temperature of cows on heat was found to be slightly higher than normal[82].

He also found that the signs of heat were more marked at the end of lactation when the cow was giving little milk; service by the bull during one heat period had no effect in intensifying the heat at the next period. The signs of heat were intensified by exercise and by putting the animal with the bull and other cows and he recommended that the bull should be kept tied up in the stalls with the cows. He also found that the signs of heat were intensified by stimulation of the vulva or clitoris by friction.

Our observations would suggest that in practice where animals are tied up during the winter they should be loosed together or with the bull at least twice a day, for in some animals the signs of heat are so slight that they would otherwise be overlooked, the presence of other animals acting in the same way as "teasers" with sheep. Also by keeping animals warm, either by shelter or exercise, the signs of heat will be more marked. The occurrence of blood from the vulva after oestrus is frequently seen even though the symptoms of heat have not been observed (because of short heat periods, see p. 18), and from this it may be estimated that the cow will be likely to come on heat again in about 16–18 days when she can be closely watched or turned out with the bull at night when not under observation.

After service by the bull it is advisable to keep a cow tied up or apart from others for frequently if this is not done she rides other cows and this is usually followed by straining and a flow of mucus with possible loss of the semen. In an excitable animal throwing a pail of cold water over her after service is said to cause the symptoms of heat to subside.

4. THE ANATOMICAL CHANGES.

Introduction.

The anatomy of the reproductive organs in the cow has been well described by Franck-Albrecht(56), Schmaltz(83) and (1), Williams(84), Sisson(85) and Montané and Bourdelle(86) so that it will be necessary only to describe the variations that take place in them.

The anatomical changes which occur in the genital organs of a cow during the various stages of the oestrous cycle have been studied by observations of specimens killed at known stages of the cycle and whenever possible measurements and weights have been taken to support these observations. Many of the changes which in the smaller animals could only be observed histologically can be seen with the naked eye in a fresh condition in the cow.

The various parts of the generative tract will be considered first separately and afterwards their relations to one another will be discussed.

Table VI gives a bird's-eye view of the changes which occur in the appearance of the various reproductive organs at different stages of the cycle in heifers of 2–3 years old.

(a) The Ovaries.

The changes occurring in the ovary during the cycle have been studied by first observing and weighing the ovaries in a fresh condition and then, after fixation in 10 per cent. formalin, by making a moderately thin central longitudinal section through them with a sharp razor. Drawings and photographs of these sections are given in Plates I–III. The sections were taken so as to cut through the largest possible diameter of the corpus luteum or largest follicle in the ovary; measurements of the maximum cross diameters (av. of 3) of these have been taken by the use of dividers and a scale.

In the cow both the follicles and corpora lutea are large and easy to measure; as a rule only one follicle ruptures at a time so that it is a very suitable species in which to study the relationships in size between the different parts of the ovary in connection with the changes occurring during the oestrous cycle.

This anatomical work would, in a small animal like the mouse, be histological and therefore has the advantage that measurements can be

PLATE I

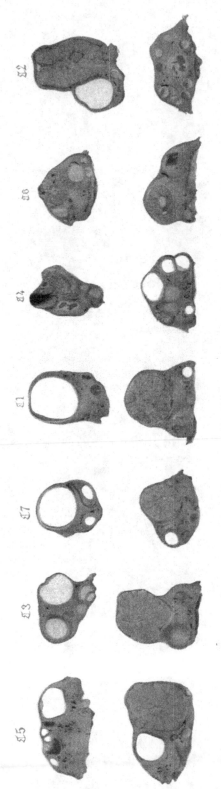

Sections of ovaries at different stages of the oestrous cycle.

Table VI. Changes observed in the reproductive organs of heifers killed at different stages of the oestrous cycle.

No. of animal:	C 5	C 3	C 7	C 1	C 4	C 6	C 2
Stage of oestrous cycle:	3 days before heat	About 14 hrs. before heat	On heat (6th hour of heat)	24 hrs. after beginning of heat	48 hrs. after beginning of heat	72 hrs. after beginning of heat	8 days after heat
Ovaries:							
Follicle	Large follicle	Very large follicle	Very large follicle	Very large follicle	Just ruptured, bleeding	Only small follicles	Large follicle
Corpus luteum	Large corpus luteum	Moderately large old corpus luteum	Moderately large old corpus luteum	Fairly large old corpus luteum	Moderately small old corpus luteum	New — small, slightly congested at point of rupture Old — moderately small	New — large Old—very small
Uterus:							
Cotyledon	Pale	Pale	Faintly pink	Slightly pink	Congested	Congested	Yellowish with brown spots
General surface	Pale	Pale	Pale. Lymph-like fluid	Pale	Pale	Congested in streaks	Pale
Cervix	Thick mucus, pale	Fluid mucus, pale	Thin clear mucus. Congested	Fluid mucus. Congested	Thick whitish mucus, pale	Clear thick mucus yellowish white at vaginal end, pale	Very thick mucus, pale
Vagina:							
Near cervix	Thick clear yellowish mucus, pale	Very fluid mucus, pale	Clear mucus with many yellow patches. Very slightly congested	Congested	Fluid mucus with streaks of fresh blood, pale	Dark brown red mucus, pale	Thick mucus, pale
Above urethra	Very slightly congested	Slightly congested	Slightly congested	Congested	Congested on surface	Pale	Congested (old)
Below urethra	Dark under the surface	Congested	Purple under the surface	Congested	Dark red under the surface	Pale	Purple under the surface

Ovulation

made without subjecting the material to the treatment necessary for histological examination.

Ovulation. Ovulation occurs spontaneously in the cow; during the course of the investigation several cases were observed in which corpora lutea were found in the ovary although no coitus had occurred at the previous heat period. This confirms the statements made on this point by Ivanoff[87], Bouin and Ancel[88] and Küpfer[89].

From Table VI it will be seen that ovulation occurs somewhere between 24 and 48 hours after the beginning of oestrus. Another cow (A 2) killed 30 hours after the commencement of oestrus had not ovulated so that the follicle probably ruptures about 40 hours after the beginning of oestrus. This finding is contrary to Nielsen's[90] statement that in the cow ovulation occurs before or at the beginning of pro-oestrum. Schmid[16] found however from rectal palpitation of the ovaries that ovulation takes place at the end of heat, *i.e.* 18–24 hours after its onset; some of the data given by him however show that ovulation had not occurred on the second day after heat began. Weber[6] killed two cows about a day before heat was due and in neither of these had ovulation taken place; another he killed on the second day of heat and in it described a large hazel-nut-sized cyst and a yellow corpus luteum of similar size which he considered had been recently formed. We believe however, that he mistook the old corpus luteum for the newly formed one (because of its size, lack of blood, and yellow colour) and the mature follicle for a cyst (since the ripe follicle about to burst enlarges) and that recent ovulation had not occurred in this animal (compare Plate I, C 1).

Krupski[91] gives a case of a cow which came on heat on the morning of April 10th and which when killed on the afternoon of April 11th—after 36 hours of heat—had not ovulated but contained a large ripe follicle in one ovary. He also cites another cow which was first seen on heat on the morning of June 17th and was killed in the afternoon of June 19th; in its ovary was found a newly formed corpus luteum. A third cow was killed 3 days after she was first seen on heat and she also had ovulated. Strodthoff[59] found from observations made on the ovaries of living cows per rectum that normally ovulation takes place at the end of heat but that if the heat is a long one it occurs in the middle of heat, *i.e.* the time of ovulation is related to the onset but not the length of the heat period.

As the normal heat period only lasts about 17 hours and ovulation does not occur until about 40 hours after the beginning of heat it follows that heat itself cannot be due to the pressure of the distended Graafian follicle on sensory nerves as Pflüger has suggested (see Marshall (92, p. 358), and that heat ends with the release of pressure which results from the rupture of the follicle. It also shows that the theory of the control of sex in cattle by service early or late in the heat period (based on ripeness of the egg) put forward by Pearl and Parshley (93) is without foundation as the egg is not shed until some time after the heat period is over.

In the cow the time of ovulation (about 40 hours after the beginning of oestrus) bears a significant relationship to the time of maximum congestion of the reproductive organs, blood appearing from the vulva in certain animals from 48 to 70 hours after the beginning of heat. The dependence of ovulation on this congestion however is doubtful for in women ovulation occurs about 8 days after menstruation has ceased. The theories as to the cause of ovulation in different species have been discussed by Marshall (92, p. 133). In the cow Zietzschmann (26) believed that the rupture of the follicle was assisted by the hyperaemia of the genitals which occurred during heat and that this acted by adding fluids to the liquor folliculi causing further swelling of the follicle and also by causing further cell proliferation in the granulosa. Joss (94) who studied the distribution of the blood vessels in the cow's ovary by means of injections concluded that ovulation occurred at a certain area in the wall of the follicle that was free from blood vessels, similar to the stigma or cleavage lines which occur in the ripe follicles of fowls (Kaupp (95)). Zschokke (40) also states that blood vessels are more numerous in the theca on the central side than on the peripheral side of the follicle in the cow, and that ovulation is sometimes prevented by a thickening of the tunica albuginea of the ovary following chronic inflammation, a cyst forming from the follicle.

The corpus luteum. The rupture of the follicle is associated with a small amount of bleeding from the orifice (Plate I, C 4) and the point of rupture in the newly formed corpus luteum can be seen as a congested patch on the surface of the ovary for the first few days after rupture. Although generally in a few days the congestion disappears the point of rupture persists throughout the life of the corpus luteum as a small knob like protrusion of luteal cells from the surface of the ovary and not until the very late stages of its involution does this disappear.

The size (average diameter) of the corpus luteum (see Table VII) has been arrived at by taking the mean of three diameters—long, including the protrusion from the surface of the ovary; one at right angles

Table VII. *Particulars of ovaries of heifers killed at different stages of the oestrous cycle.*

Stage of cycle:	3 days before heat	14 hrs. before heat due	Oestrus 6th hour of heat	24 hrs. after beginning of heat	48 hrs. after beginning of heat	72 hrs. after beginning of heat	8 days after heat
No. of heifer:	C 5	C 3	C 7	C 1	C 4	C 6	C 2
Weight of ovaries (gm.):							
With corpus luteum before ovulation and follicle after it	8·5	5·2	3·8	4·8	3·1	3·2	3·5
With follicle before ovulation and with new corpus luteum after it	4·2	3·6	5·5	4·9	4·1*	2·6	9·4
Size of corpora lutea (diam. cm.):							
Of new generation	—	—	—	—	0·8	1·1	1·9
Of last generation	1·9	1·7	1·3	1·6	0·8	1·2	—
Of previous generation	0·6	—	—	—	—	0·5	—
Size of follicles (diam. cm.):							
To ripen at oestrus	1·1	1·2	1·3	1·5	—	—	—
To ripen at next oestrus	0·8	0·8	0·9	0·4	0·9	0·7	1·2
To ripen at subsequent oestrus	—	0·7	0·7	—	0·6	0·5	0·4
Colour of corpus luteum:							
New generation	—	—	—	—	White—blood centre	Cream—blood at apex	—
Last generation	Yellow	—	Bright yellow	—	Yellow	Canary yellow	—
Previous generation	Orange red	—	—	—	—	Orange red	—
Form of corpus luteum:							
New generation	—	—	—	—	Round—blood clot in centre	Round—central cavity with albuminous fluid	Elliptical loose centre
Last generation	Ovoid—dense	Nearly round—connective tissue plug in centre	Round—dense	Round—strands of connective tissue in centre	Flattened—dense	Rather flattened cystic in centre	Flattened
Previous generation	Flattened Triangular	—	—	—	—	Flattened	—

Ovulation

* Both new and old corpora lutea in same ovary.

to this and a third at 45° to it—in two surfaces of the corpus luteum exposed by the section.

The newly formed corpus luteum (48 hours after the beginning of oestrus) is only about half as large (0·8 cm.) as the follicle from which it is formed but it enlarges rapidly and attains full size (1·9 cm.) at or before mid-cycle. The heifer C 2 which was killed at 8 days after oestrus had a normal cycle of 17 days and so represents the mid-cycle size. This development is maintained or slightly exceeded until three days before the next heat period when it begins to diminish in size rapidly. Old corpora lutea to the third and fourth generation can be traced in the ovaries and they eventually degenerate into patches of lipochrome pigment in the stroma of the ovary. The persistence of the old corpora lutea in the ovaries of cows and their transformation into lipochrome patches or corpora rubra have been described by Zietzschmann[26] and also by Küpfer[89] who has made a very thorough study of the anatomy of the ovary and especially the corpus luteum in the cow. Küpfer gives a chart showing the curves of development and regression of successive generations of corpora lutea in the ovaries of cows and shows that the corpus luteum reaches its maximum development at 10–11 days and then begins to decrease in size. Schmid[16] states that the corpus luteum attains full size in about 4–6 days and remains large until about 6–8 days before the next ovulation; the sizes of the corpora lutea being as follows: 1 day after heat, 0·9 cm.; 3 days after, 1·7 cm.; 8–10 days, 1·6 cm.; 11 days, 1·9 cm. and 14 days after, 2·0 cm.

Strodthoff[59] found from observations per rectum that the corpus luteum can be felt 3 days after heat; at 5 days it is well developed; at 12 days full size is reached and at the time of the next heat it can still be felt in the ovary.

The colour of the corpus luteum changes during its life. When first formed it is pale but stained with blood and becomes cream in about three days; it then changes to a bright canary yellow, turning reddish orange when old and as a "corpus rubrum" is frequently bright scarlet but sometimes brownish. Williams[84, p. 17] states that the colour of the corpus luteum of menstruation is chocolate but the corpus luteum of pregnancy is orange, but this is probably due to his having examined early blood stained stages of the corpus luteum of menstruation only.

The reason for the changes in the colour of the corpus luteum is probably to be found first in the blood content of the centre in the early stages and secondly in the rarity and then formation and concentration of luteal pigment which is associated with the lipoid droplets deposited in the luteal cells. This pigment as Palmer and Eckles[96] found with butter-fat is probably derived from the chlorophyl of plants and as the lipoids are deposited in the corpus luteum it is brought in attached to them. During the process of absorption of the luteal cells on involution of the corpus luteum the pigments would appear not to be reabsorbed with the lipoid but, remaining in the corpus luteum, become concentrated and give rise to the deep colour. Since certain vitamines are frequently associated with the colour of plants it may be that they also when once deposited in the body are not so easily transferred as the substances to which they are attached.

Escher[97] investigated the chemical composition of luteal pigment and found it to be practically identical with the carrotine obtained from green leaves and very similar to that of the yellow pigment contained in body fat and the yolk of hens' eggs.

It is notable that the colour of the corpus luteum soon fades by oxidation when it is kept in formalin in the light; the colour is to some extent brought back however if the specimens are treated with a reducing agent. The specimens shown in Plates I–III were fixed in formalin, passed rapidly through alcohols (for the pigment is dissolved, along with the lipoid, by 75 and 95 per cent. alcohol) and were cleared in a solution of glycerine (50 c.c.), distilled water (50 c.c.) and potassium acetate (2 gm.). Palmer and Kempster[98] found that in fowls with yellow skins due to accumulation of this pigment oxidation occurs causing loss of colour when incoming supplies are devoted to egg production.

The prevalence of the bright scarlet colour of the old corpora lutea in fattened animals (Marshall and Peel[53]) may possibly be due to the accumulation of fat causing the oxidation processes occurring in the tissues to be at a minimum. How far the colour of the corpus luteum depends on breed and food conditions, just as the colour of body and milk fat does, has not been determined. Corner[99] found that in pigs, where the body fat is normally white, the colour of the corpus luteum is not yellow but a light pinkish grey and only turns yellow when old. Küpfer's[89] illustrations also suggest that the corpora lutea are of deeper colour when the animal has been grass fed (killed in June) than when stall fed (killed in February).

The corpus luteum when first formed (48 hours after the commencement of heat) contains a central blood clot (Plate I, C 4) surrounded on all sides except the point of rupture by luteal tissue. Delestre[100] doubted whether bleeding occurred in the ruptured follicle of the cow but he based his conclusions on corpora lutea which were several days old.

Zietzschmann (26) however states that extravasation of blood into the tissues occurs at the point of rupture of the follicle and possibly also into the cavity. Küpfer (89) has figured freshly formed corpora lutea in cows showing an exudation of blood at the point of rupture. Zschokke (101) states that bleeding may, but does not always occur on rupture of the follicle in the cow. Krupski (91) found blood coagula in the freshly ruptured follicles of cows and believed that this prevented bleeding from the orifice; he found that occasionally the blood clot persisted and was seen in the corpus luteum 10 days after the last heat period. We have also noticed the presence of a small blood clot in the corpus luteum of a cow killed about mid-cycle (Plate II, 40).

In the next stage of the corpus luteum (Plate I, C 6—72 hours after commencement of heat) the central cavity becomes filled with a glairy albuminous fluid similar to the liquor folliculi, the blood clot being still visible at the point of rupture. This stage is well shown in a series of cows from slaughter-houses (Plate II and Table XI) which were grouped into stages of the cycle by comparison with the series of heifers (Plate I and Table VI) but whose exact period was not known.

This stage has been described in the cow by Delestre (100); 4 out of 12 corpora lutea of menstruation obtained at random contained a central cavity filled with fluid and he considered it to be a normal condition due to the redistention of the follicular cavity with fluid. Meyer (72) who found cystic corpora lutea in women concluded that their effect on the uterine mucosa is not different from that of other corpora lutea.

Küpfer (89) has described in cows a small central cavity in the corpora lutea of 4–5 days old and also larger cavities at 8–12 days old which he believed to be luteal cysts.

Schmid (16) describes in a cow killed a day after heat a corpus luteum found in a not yet ruptured follicle.

Zschokke (40) states that the newly formed corpus luteum of the cow has a central cavity which is filled with fluid or a blood clot.

Krupski (91) states that, in the stage following the blood coagulum, the cavity of the corpus luteum becomes filled with yellow serum; he concluded that the central cavity becomes filled with much fluid if the point of rupture closes before the cavity is filled with luteal cells. These abnormal corpora lutea he states undergo regressive changes similar to those occurring in the normal ones. Williams (84, p. 155) considers that this form of the corpus luteum exists only in cases of sterility and is due to the cystic degeneration, the luteal substance disappearing and a cyst forming.

There can be little doubt however that the stage with fluid in the central cavity is a normal one in the life of the corpus luteum. In other species a similar phase in the formation of the corpus luteum has been found; Sobotta (102) has described and figured it in the rabbit and Long and Evans (32) in the rat. Robinson (51) has shown that in ferrets the follicles are normally redistended with fluid after ovulation.

In Plate II, cows Nos. 32, 4, 34 and 21, the corpora lutea are seen to contain a large development of the central albuminous fluid. From appearances it would seem that this fluid is the same as that of the liquor folliculi and that after the egg is shed the granulosa cells still go on producing this fluid; it is not until the capillaries of the corpus luteum become organised that this substance is absorbed into the blood. It is conceivable that under certain abnormal conditions the reabsorption of the fluid may not occur but give rise to cysts. It may be that the production and absorption of this substance supplies the active internal secretion of the corpus luteum. It has been shown that the presence of ripe follicles in the ovaries has much the same but a smaller effect on the uterus and mammary gland than the presence of the active corpus luteum (Nielsen (90)). The same substance—liquor folliculi—would appear to be produced by both from the same cells—granulosa and then luteal cells—and the reason why the effect is so much greater with the active corpus luteum may be that here the cells are in close apposition to the blood capillaries while in the follicle the basilar membrane and a dense layer of granulosa cells separates the fluid from the capillaries deep in the theca (see further p. 80 below).

By 8 days after the beginning of heat, if not before (Plate I, C 2) the whole of this fluid is normally absorbed and the central cavity is packed with a branching plug of white connective tissue. This plug of connective tissue gets smaller and more condensed as the life of the corpus luteum proceeds but its persistence is very marked and it can be seen in very late stages of the corpus luteum, and often in lipochrome patches, as a small central white plug. In some cases the absorption of the central fluid does not take place completely; Küpfer (89) describes two cases which he considered were luteal cysts with a central cavity filled with fluid 8 and 12 days after the last heat period. Delestre (100) has noted a small amount present in a cow pregnant $2\frac{1}{2}$ months and he describes the connection tissue septa which divide up the corpus luteum as converging to it and surrounding it. We have seen one such case in

PLATE II

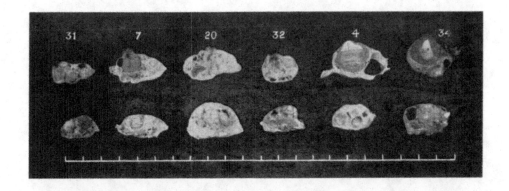

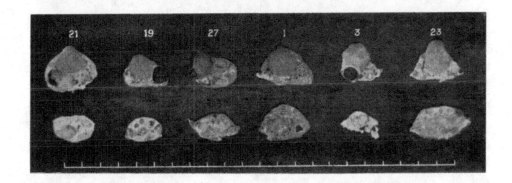

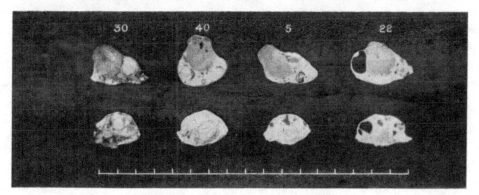

Ovaries of cows obtained from slaughterhouses.

which a corpus luteum of 21 days old contained a centre of albuminous fluid (Plate I, C 6).

Küpfer[89] states that in cases of twin ovulations in the cow both corpora lutea go through the various stages of development together whether both or only one embryo develops, which corresponds to the development of these bodies in pigs and rabbits under similar conditions (Hammond[103]). In cases of twin ovulation Küpfer states that each corpus luteum is slightly smaller than it would be normally where only one developed. Fraenkel[42] has shown that in the rabbit the number of corpora lutea can be reduced to more than a half less than the number of embryos before the normal development of the latter is affected.

The shape of the corpus luteum (see Table VII) is mainly dependent on pressure conditions in the ovary. It is round in the early stages but at the time of maximum development tends to lengthen and grow out of the ovary as pressure is less on this side. With the onset of the regression phase it again becomes round and in the later stages of the phase becomes flattened, or triangular if follicles develop on two sides of it.

The Graafian follicle. The measurements of the Graafian follicle have been taken in the same way as those of the corpus luteum. In all cases observed (just over 60) only one follicle ripened at each period. Curot[22] states that twin births occur in the cow once in every 80 cases. Küpfer[89] has described several cases of twin ovulations and one of three follicles rupturing at the same time. It may be of more frequent occurrence in some breeds than others. Williams[84, p. 170] states that twin pregnancies are most common in Friesians and rarest in Jerseys.

The measurements given in Table VII show that the enlargement of the follicle that is to ripen begins just before mid-period and that a very gradual increase in size occurs until just after heat when a sudden enlargement takes place; possibly this enlargement is associated with the congestion of the genital tract at this time and with engorgement of the thecal blood vessels (but see p. 35) and culminates in rupture. Robinson[51] found that there was a marked swelling of the follicle in ferrets just before ovulation. Zschokke[101] found that the ripe follicle in the cow measures 1–1·2 cm. in diameter and Krupski[91] found that it was 1·3 cm. Schmid[16] found the follicles to be rather larger (possibly a breed difference) and says that at 3 days after heat they were 0·7 cm., at 8–10 days 1·2 cm. and at 14 days 1·6 cm.

A second follicle generally follows the first in growth at a lower level and as a rule several others of about 0·5 cm. diameter exist in the ovary.

The follicle attains almost mature size (excluding the maturative enlargement) at 4–5 months old in calves which have not reached the age of puberty (Plate III, 1, Nos. 11, 12, 24) as will be seen from Table VIII;

Table VIII. *Weights and sizes of reproductive organs of calves* *(before first ovulation).*

Age about (months)	No. of animal	Ovaries weight in gm.		Follicles diameter in cm.		Uterus weight in gm.	Udder weight in gm.
		Largest	Smallest	Largest	Second largest		
2	2	1·5	1·5	0·65	0·63	22·5	220
3–4	8	1·4	0·8	0·8	0·65	61·0	640
4–5	24	2·2	1·2	1·1	0·4	47·0	—
5	12	2·5	1·2	1·0	0·7	40·0	440
6	11	2·9	1·4	1·2	1·0*	136·0	760
					0·8†		
6	25	3·3	2·9	1·0	0·9	53·0	640
	Average	2·3	1·5	0·96	0·71	59·9	540

* In same ovary.   † In other ovary.

the age of puberty is probably reached about 9 months old. Heitz[104] found that follicles quite as large as ripe follicles occurred in the ovaries of calves 5–12 weeks old, but these did not rupture or form corpora lutea although some became atrophic or formed cysts. Of 75 ovaries of calves between 5–12 weeks old 80 per cent. contained a follicle of over 0·3 cm. in diameter and in four cases a follicle of 1·3 cm. diameter was found. These results have been confirmed by Käppeli[105].

No explanation can be given as to why the follicles should not ripen at 5 months but wait until 9 months before doing so. A certain amount of information however may be obtained by comparing the weights of the two ovaries of an animal (see Table IX). In a calf of two months,

Table IX. *Relative weights of the two ovaries in calves and heifers.*

(a) Calves before ovulation.

Approximate age—months	2	3–4	4–5	5	6	6
Ratio $\frac{\text{With large follicle}}{\text{Other}} = 100$	100	175	183	208	207	114

(b) Heifers during oestrous cycle.

Period of cycle	3 days before heat	14 hours before heat due	6th hour of heat	24 hours after beginning of heat	Ovulation	48 hours after beginning of heat	72 hours after beginning of heat	8 days after heat
No. of animal	C 5	C 3	C 7	C 1		C 4	C 6	C 2
Sum weight of both ovaries (gm.)	12·7	8·8	9·3	9·7		7·2	5·8	12·9
Ratio $\frac{\text{Ovary with corpus luteum}}{\text{With follicle}} = 100$	202	144	69	92		132*	81	269

* Both new and old corpora lutea in same ovary.

PLATE III

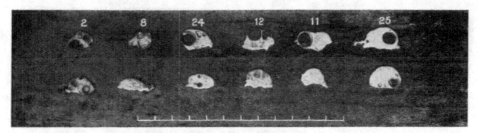

1. Ovaries of calves.

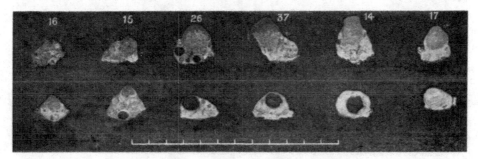

2. Ovaries of heifers from slaughterhouses.

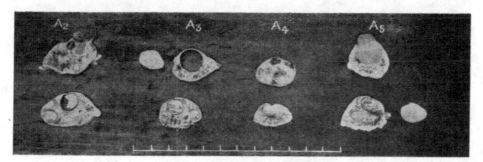

3. Ovaries of experimental cows.

before the follicles have begun to develop at all, both ovaries are equal in weight; as the follicle enlarges the ovary containing it increases in weight and its ratio to the other ovary increases up to about 5 months old but then instead of developing further the next largest follicle (usually situated in the other ovary) begins to grow so that the ratio between the two ovaries is again decreased. It would appear that the special nutriment available for the growth of the follicles is used to nourish a series and keep them all going before the largest one can mature. Küpfer(89) has described the ovaries of heifers before puberty and found in them no corpora lutea but only moderate and large sized follicles; in 10 heifers killed before puberty he found that the weight of the ovary was roughly proportional to the number of follicles that could be observed with the naked eye so that growth in weight of the ovary at this stage was dependent on growth of the follicles.

When the maturation of the follicle is considered in relation to the ovary as a whole in animals which have reached the age of puberty the effect of the corpus luteum has to be considered. Experimental evidence (see p. 23) has shown that this body hinders ripening of the follicle. Diagram I which shows the relative sizes of the different parts of the ovary at various stages of the cycle confirms this conclusion. In this diagram the curves of growth of the different parts have been smoothed as differences of age, size, time of year, etc. cause slight variations and sufficient material was not available to eliminate these by averages. It will be seen that while the corpus luteum remains large the increase in size of the follicle is only very gradual and that as soon as the corpus luteum begins to atrophy, which it does rapidly, the follicle hastens to mature and ovulation occurs. It would seem that while the corpus luteum is active it has first call on the special nourishment supplied to the ovary but when it begins to atrophy the supply is then available for the largest follicle and its growth is hastened.

The curve given in this diagram for the growth of the corpus luteum in the cow does not agree in detail with that given by Küpfer(89) who concluded that hypertrophy of the corpus luteum continues only to the 10th–11th days and that the 11th–21st days are occupied by degenerative changes. His curve is a peaked one with the maximum at 10–11 days rather than flattened as shown in Diagram I. He agrees however that degeneration of the corpus luteum is rapid at the time of ripening of the new follicle. The weights and measurements of corpora lutea on which he bases his curve were however obtained from animals of different breeds and ages and it is shown in this paper that both these factors

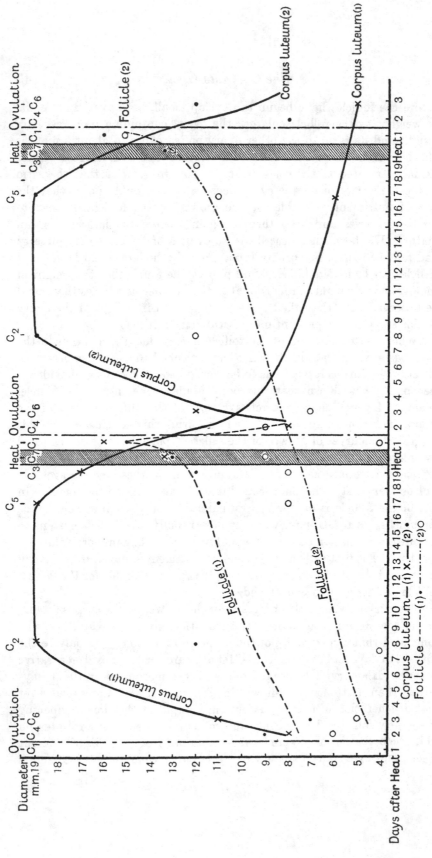

Diagram I. Curves of growth of follicle and corpus luteum during the oestrous cycle.

affect the size of the corpus luteum; in our experience the corpus lutuem is slightly larger in old animals than in young and also in Jerseys as compared to Shorthorns. At 10 days old his corpora lutea had an average diameter of 2·6 cm. (which corresponds to our 1·9 cm. at 8 days as this animal was one with a 17 day cycle) and at 17 days was 2·0 cm. (corresponding to our 1·9 cm., 3 days before heat); the average size at the time of the next heat (20th day) was 1·5 cm. which corresponds more or less with our results. Krupski[91], who killed three cows about a day before the heat period was due, found the corpus luteum to have a maximum diameter of 1·5–1·7 cm., the ripe follicle varying from 0·7–1·2 cm. in diameter.

It must be concluded that growth of the follicle does not depend alone on material derived from other follicles as Robinson[51] supposes but on some substance in the blood utilised by both follicle and corpus luteum. The large amount of follicular atrophy that occurs when a corpus luteum is present in the ovary (see p. 128) supports this view. Loeb[43] however found in guinea-pigs that the corpus luteum does not prevent ripening of the follicle but prevents ovulation; but the effect of the corpus luteum on the growth of the follicle (Sandes[106]) appears to vary in different species (see p. 27).

Corner[107] found in sows that not more than 2–3 days before oestrus there was a rapid enlargement of the follicles corresponding to a decrease in size of the corpora lutea which commenced about 5 days before the heat period.

Ovulation is not prevented by the presence of a corpus luteum in the ovary but by its state of development, that is whether it is utilising the common basis of nourishment of follicle and corpus luteum ("Generative ferment") or whether it has become atrophic.

Zietzschmann[26] has pointed out that in the cow the corpus luteum cannot prevent ovulation mechanically as the follicle may often ripen in the other ovary.

The size of the ovary. Käppeli[105] found that in cows the state of nutrition and early maturity of the breed affected the growth of the ovaries; while during the first month of life the ovary grows at a faster rate than the rest of the body, yet with the onset of puberty the rate of ovarian growth is less than that of the body as a whole.

A study of the weights of ovaries at different stages of the cycle (Table VII) shows that they are influenced very largely by the presence of the corpus luteum which (in heifers of about $2\frac{1}{2}$ years old) when at its maximum development may weigh nearly twice as much as the rest of the ovary (see difference in weight between two ovaries of heifers C 2 and C 5, Table VII). Consequently the weight of both ovaries together varies with the stage of the cycle, being greatest from 8 days after to 3 days before heat and smallest just after ovulation.

The ratio in weight of one ovary to the other is interesting (Table IX(*b*)). Three days before heat the weight of the ovary with the corpus luteum is twice that of the other (with the follicle) but during heat it is only

Table X. *Weights and sizes of reproductive organs of heifers (after 1st ovulation and before 1st calf).*

Approximate stage of oestrous cycle	No. of animal	Approx. age of animal Yrs.	Mths.	Ovaries: weight gm. With last corpus luteum	With largest follicle	Corpus luteum: diameter cm. New	Old	Follicle: diameter cm. Largest	Second largest	Uterus weight in gm.	Udder weight in gm.
Just before heat	14	2	4	8·3	5·5	—	1·7	1·3	0·4	340	1600
1 day after heat	17	1	6	4·5	3·6	—	1·6	1·55	0·4	280	1420
Ovulation											
2 days after heat	16	1	11	3·6	3·2	0·75	1·0	0·6	0·6	280	2380
3 ,,	15	3	0	4·0	5·8	1·3	0·9	0·7	0·6	620	2650
12 ,,	26	1	9	8·3	3·9	2·1	—	1·2	0·9	200	1600
17 ,,	37	2	0	9·2	4·8	2·1	—	1·3	0·6	255	—
Average:				6·3	4·5	1·6	0·9	1·1	0·6	329	1930

Table XI. *Weights and sizes of reproductive organs of cows (after 1st calf).*

Approx. stage of oestrous cycle days after heat)	No. of animal	Approx. age of animal	Ovaries: weight gm. With last formed corpus luteum	With next ripening follicle	Corpus luteum: diameter cm. New	Old	Follicle: diameter cm. Largest	Second largest	Uterus weight in gm.	Udder weight in gm.
2	31	?	5·0	3·5	0·9	1·1	0·9	0·8	400	1680
3	20	Old	15·5	10·4	1·2	0·9	0·6	0·6	720	4660
3	32	?	6·0	4·5	1·3	—	0·9	0·9	480	3100
4	7	8 years	?	7·5	1·4	0·6	1·0	0·6	840	8980
5	4*	12 ,, (calved 6 mths.)	16·0	5·0	2·3	0·6	1·2	0·8	900	7820
5	34*	7 years (calved 4 mths.)	12·7	7·4	2·1	0·9	1·1	—	1000	12000 (diseased)
6	21	Old	16·7	7·8	1·9	0·7	1·4	1·0	840	7120
8	27	,,	11·1	4·5	1·8	—	1·1	0·4	520	2480
10	1	Over 3½ years	15·5	10·5	2·0	—	0·8	—	1280	5760
10	23	5 years	11·3	8·9	2·0	—	1·1	1·0	720	3900
10	40†	13 ,, (calved 2 mths.)	15·6	8·9	2·1	—	1·4	0·9	610	12690 (diseased)
12	30	Old	10·0	7·0	2·1	—	1·2	1·0	960	7840
12	5	?	12·0	7·0	2·0	—	1·2	0·7	600	5020
15	19	?	9·5	4·3	1·7	—	1·2	0·6	340	—
15	3	Over 3½ years	7·5	3·0	1·8	—	1·2	1·1	460	4880
15	22	Young	16·6	7·0	2·2	—	1·3	0·9	760	3040
Average:			12·1	6·7	1·8	0·8	1·1	0·8	714	5100

* Shorthorn.　　† Jersey—others all Shorthorn cross breeds.

about 75 per cent. of its weight, this being caused by the atrophy of the corpus luteum in the one and growth of the follicle in the other. Excluding the case where ovulation occurred in the same ovary that contained the old corpus luteum, the ratio between the ovaries is 80 per cent. soon after ovulation and this rises to almost three times the size 8 days after the beginning of oestrus, due to the formation of the new corpus luteum.

Since the body weight of the cow shows no great change during the different stages of the oestrous cycle the significance of the great changes in ovarian weight is seen when the ratio of ovary to body weight is compared. The ovary should therefore exert its maximum influence over body metabolism at mid-period. This ratio of ovary weight to body weight may have a practical importance, as will be pointed out below, in determining the milking qualities of an animal.

If in the foregoing tables (VII–XI) a comparison is made of the sizes and weights of the different parts of the ovary in calves, heifers and cows (Plates I–III) it will be seen that (with the exception that the maturative increase of the follicle does not occur in calves) the size of the mature follicle is approximately the same whatever the age. The maximum size of the corpus luteum however appears to be slightly larger in cows than in heifers and may be the cause of their rather longer oestrous cycles. This small increase in size of the corpus luteum however will not account for the large differences in weight between the ovaries of cows and heifers.

There is only a little increase in the weight of the ovary (apart from the corpus luteum) between a calf of 6 months (ovary weight 3·1 gm.) and a heifer of 2 years (ovary weight 3·4 gm.) but in old cows (ovary weight up to 10 gm.) the increase in weight is very marked, the ovary without the corpus luteum frequently weighing from 7 to 10 gm. Schmid (16) has given the weights of ovaries of heifers and cows during the cycle and these show that the cow's ovary is very much heavier than the heifer's. The photographs given in Plates I–III show that the cause of increased ovarian weight is increased growth of ovarian stroma. It will be noticed that this increase takes place not between 6 months and 2 years when puberty occurs, and so is not due (for proof see p. 83) to any growth of interstitial gland, but increases with the breeding life of the animal and probably consists of connective tissue growth derived largely from the central plug and surrounding connective tissue of the successive corpora lutea. This would explain why fibrous ovaries are so common in old females that have not been allowed to breed, since the presence of a corpus luteum of pregnancy would in the cow prevent the formation of about 13 plugs of connective tissue in the ovary. Blair Bell[108] states that as a rule a woman menstruates regularly for about 30 years; in the case of a multipara however, the periods of gestation do not count, as it were, and she continues to menstruate by so much the longer. Küpfer[89] has pointed out that the ovaries of heifers have as a rule a larger proportion of their follicles on the surface of the ovary than have cows and this may be caused by the greater development of connective tissue in the ovary of the cow. He has also shown that ovarian growth does not cease at puberty but that there is no direct relation between the age and the number of follicles in the ovary. Käppeli[105] has shown by counting the follicles in serial sections of the ovary that whereas on the average the number of follicles in the ovaries of calves of 3 months old is approximately 75,000 it is reduced to 21,000 in heifers of $1\frac{1}{2}$–3 years old and is as low as 2500 in old cows.

Alternating action of the ovaries. Strodthoff[59] from clinical observation, states that ovulation from the same ovary as that in which the last ovulation occurred is not frequent. Küpfer[89] states that in the cow usually first the right and then the left ovary ovulates, but that this does not always occur and in some animals only one ovary is functional. He found that the right ovary has a tendency to function more than the left, the average weight of the right ovary being 7·15 gm. and the left 5·54 gm. in 45 cases in non-pregnant heifers; in 24 cases of non-pregnant cows the right ovary averaged 12·33 gm. and the left 10·70 gm. This difference in weight was due mainly to differences in the number of corpora lutea or follicles found. Heitz[104] who weighed the ovaries of 75 calves between the ages of 5 and 12 weeks found that the right ovary averaged 1·62 gm. and the left 1·65 gm. Simon[109] who weighed the ovaries of 95 castrated cows (many with cysts) found that the average weight of the right was 25·4 gm. and the left 20·3 gm. Schmid[16] from 100 cows found the right ovary weighed 8·61 gm. and the left 6·51 gm. Käppeli[105] from an examination of the ovaries of 93 calves, heifers and cows concluded that on the whole the right ovary is heavier than the left and that the weight of the ovary is mainly dependent on the growth of the follicles or corpora lutea and that this is the reason why one ovary is heavier than the other. Krupski[91] also found that the ripe follicle or fresh corpus luteum occurred more frequently in the right ovary than in the left.

In 65 per cent. of the animals we examined the follicle ripened in the opposite ovary to that in which the previous ovulation occurred.

No doubt the pressure of an active corpus luteum would tend to lower the chances of nutrition being favourable for development in the same ovary since the cells of each are of the same origin and consequently probably draw on the same sources of nourishment. Sandes[106] noticed that during pregnancy the follicles atrophied in ever widening circles round the corpus luteum but this he attributed to mechanical pressure or to an internal secretion of the corpus luteum. It is notable that in cases where ovulation occurs in the same ovary as the previous one the onset of heat is slightly delayed but the effect on the duration of heat is not quite so marked.

Except for this action of the corpus luteum in rendering the chances of a follicle ripening in the same ovary less favourable there would seem to be no fundamental influence at work to regulate the alternating action of the ovaries and except for the influence of the corpus luteum the chances of a follicle ripening in the same or other ovary would be equal with a tendency, as shown by the results of various authors quoted above, for it to occur slightly more frequently in the right ovary. Possibly the presence of the rumen on the left side of the body hinders blood flow to this side as compared with the right.

(b) The Uterus.

The anatomy and blood supply to the uterus has been described by Zieger[110] who states that the cross section is usually oval when at rest but on erection it becomes round and the horns are shortened.

The anatomical changes have been investigated by weighing the uterus, Fallopian tubes and cervix together; these were separated from the vagina by cutting round its wall just posterior to the external os, the ovaries and fatty tissue adhering to the Fallopian tubes being first cut away. Afterwards the body and one horn of the uterus were cut open and their mucous membrane examined. Table VI shows the changes which were observed.

It will be seen from this table that the mucous membrane of the uterus does not become congested until 2–3 days after heat has commenced. The first appearance of congestion occurs in the cotyledons which are slightly pink just after heat ends (24 hours after the beginning of heat) at which time the general surface of the uterus is still pale. Forty-eight hours after heat commences the cotyledons are quite congested and red but the general surface does not show any change and it is not until 72 hours after the beginning of heat that the general surface of the uterus becomes bright red and congested. Even at this

H

4

time there is not congestion of the entire surface and the red patches appear in streaks between the cotyledons. By 8 days after the commencement of heat the congestion has disappeared leaving only a yellowish brown coloration where it has been most intense, namely on the cotyledons. At other periods of the cycle before the onset of heat the general uterine mucosa and the cotyledons are quite pale. During oestrus a thin lymph-like fluid is seen on the general surface of the mucosa, this being probably a secretion of the uterine glands. This fluid in the uterine cavity at the time of oestrus has been observed by Long and Evans[32] in the rat and by Gerlinger[111] in other species; he believes that it serves the function of allowing the spermatozoa to swim freely and so ascend the female generative tract.

Emrys-Roberts[112] found a watery secretion tinged with blood in the uterine horn of a cow killed in a pro-oestral condition but we have never seen blood at this stage of the cycle. He states that the secretion in the pro-oestral condition was watery in comparison with the somewhat mucoid condition in the resting state. Schmid[16] states that in cows there is bleeding into the cavity of the uterus; he found blood and slime in the uterus of a heifer that was on heat the day before.

Weber[6] took two cows that had previously bled after their heat periods and killed one a day before heat was due, in this the uterus was pale and contained a little clear fluid; the other he killed a day after heat and found that the mucosa of the uterus was swollen, yellowish and darker than normal, no blood but only clear mucus being present in the uterine cavity.

Zietzschmann[26] states that in heifers during heat the glandular secretion reaches a maximum and extravasation of blood occurs in the uterus after the high point of heat has been reached; blood accumulates in the subepithelial mucosa and here and there penetrates into the uterine cavity and colours the mucus.

Utz[113] found in a cow at the time of heat the cavity of the uterus contained secretion stained with blood and that the horns of the uterus were more congested than the body and that the cotyledons showed the greatest congestion which he considered was the source of the blood of the uterus.

Küpfer[89] figures one case of a cow with a newly ruptured follicle in the ovaries in which the cotyledons of the uterus were very congested; he states that uterine bleeding does not occur in all cases where there is a fresh corpus luteum in the ovary. Krupski[91] and [114] states that bleeding occurs in cows at the end of heat. He found blood-stained slime in the uterine cavity and blood coagula on the cotyledons but states that they vary considerably in different individuals, bleeding being much more frequent in heifers than in cows (see p. 57). In old animals when the congestion was not so marked it was frequently confined only to the side on which ovulation occurred, but this we are unable to confirm. In a cow killed 36 hours after heat had commenced he found that the uterus was swollen and contained clear mucus in the cavity. In a cow which was on heat on the morning of January 17th and was killed in the afternoon of January 19th he found the uterine mucosa, and especially the cotyledons of both sides, congested. In another cow killed 3 days after heat the uterus was normal, but one killed 4 days after heat showed blood coagula on the cotyledons of one horn of the uterus only.

The relation of the time of congestion of the uterus to the time of the appearance of blood from the vulva is a matter of some importance in determining the origin of the latter for at 72 hours after the commencement of heat the blood vessels of the uterus were still congested and had not broken down while in the majority of cases (see Table XII) bleeding occurs from the vulva from 48 to 74 hours after the beginning of oestrus. This together with other findings (see p. 57) suggests that the external bleeding in the cow is derived mainly from the vagina and not from the uterus. The occurrence of the brown pigment in the cotyledons after the time of congestion points to the absorption of the extravasated blood *in situ* (as Marshall[54] found in the sheep) rather than its evacuation by bleeding. It might be imagined that the pro-oestrous bleeding followed by oestrus such as occurs in the bitch is telescoped in the cow owing to the quickness with which cycles follow one another, but we believe that this is not so for bleeding in the cow rarely occurs before 48 hours after the commencement of heat at a time when desire has ceased and ovulation has occurred.

The weights of the uterus in different animals are shown in Tables VIII, X and XI. With calves (Table VIII) the variation in weight is from 23 gm. at 2 months old to 136 gm. at 6 months old, the average at 4 months being 60 gm.

In virgin heifers (Table X) the weight varied from 200 gm. at 1 year 9 months old to 620 gm. at 3 years old, the average at 2 years 3 months old being 330 gm.

In cows which had produced a calf (Table XI) and which varied considerably in age the weight of the uterus varied from 400 gm. to 1280 gm., the average being 714 gm. Hess[41] states that in Simmental cows it averages 700–800 gm. Sommer[115] found that while in virgin animals both horns of the uterus were equal in size, in cows which had produced calves one side (usually the right) was larger than the other owing to the development which had taken place in pregnancy (pregnancy being more frequent on the right side).

Within each of these three groups—calves, heifers and cows—there is much less variation (except in cows) than there is between the different groups, which is to be expected, for it has been shown that the development of the uterus depends on the ovaries, ovulation and the formation of the corpus luteum being responsible for the rise from 60 gm. in calves to 330 gm. in heifers. The further rise to 714 gm. in cows is probably partly due to the further activity of the corpus luteum developed in pregnancy and partly as a reaction to the growth of the foetal membranes.

The weight changes of the uterus during pregnancy will be discussed later (p. 129).

Mayer[116] who investigated the effect of the ovaries on the growth of the uterus in rabbits concluded that its growth was not dependent on the ovaries during foetal and infantile life but was so after puberty. He based these conclusions on the size to which the organ involuted when the ovaries were removed at different ages.

Unfortunately the weight of the uterus in all the heifers killed at different stages of the oestrous cycle were not taken but in those of which records exist the differences were small. The great variation in the weight of the uterus in cows may be due in part to the sub- or hyper-involution of the uterus after pregnancy.

Servatius[117] however found that involution normally takes place quickly, the uterus at the 4th day post-partum being reduced to half its size and by the 8th day to a third and by the 14th day it is practically normal in size.

Sommer[115], who investigated the involution of the uterus in the cow, found that about an hour after birth it weighed 10,000 gm., two days post-partum about 6700 gm., five days post-partum 3400 gm. and after four weeks about 576 gm., the normal weight. The weight then usually diminished gradually until 6–7 weeks after parturition when it may be under normal weight (super-involuted), a condition which is associated with lactation in many species (Loeb[118], Hammond and Marshall[24]); the weight of the virgin uterus he found to be 220 gm.; Hilty[119] also gives the weight of the uterus under different conditions; in a heifer of 17 months it was 128 gm., in a cow pregnant 6 weeks 890 gm. and at the 16th, 21st, 29th and 43rd day after parturition it weighed 1250 gm., 1150 gm., 980 gm. and 760 gm. respectively.

Albrechtsen[120] from clinical observations concluded that involution is finished in 3 to 4 weeks when the cow normally comes on heat again; he also states that heat is often delayed after this time and this is caused by incomplete involution of the uterus. We however believe that delay both in the onset of heat and in the involution of the uterus in such cases is caused by the retarded atrophy of the corpus luteum.

The involution of the cotyledons after parturition Sommer also found to be very rapid; the average weight of a cotyledon one hour after birth was 70 gm., 2 days post-partum 26 gm. and by 5 days after they had become quite small, being practically back to their normal size by 9 days after parturition.

The maximum size of the cotyledon (mm.) in the cow attained under different sexual conditions was found by Hilty[119] to be as follows:

	Length	Breadth	Height
Virgin heifer 17 months	6	8	3
Cow pregnant 6 weeks	9	4	3
,, 9 ,,	18	9	6
,, 13 ,,	22	13	16
,, 15 ,,	31	19	11
,, 20 ,,	59	38	19
,, 26 ,,	78	34	26
Cow 3 days post-partum	74	50	17
,, 7 ,,	45	36	13
,, 15 ,,	25	13	6
,, 16 ,,	13	12	4
,, 21 ,,	13	11	4
,, 29 ,,	10	6	4
,, 43 ,,	10	3	4

Thus it will be seen that during pregnancy growth is made by the uterus which is never entirely lost again and since the cotyledons regain their normal size the difference probably lies in the muscular development and its associated connective tissue and blood vessels.

Sommer gives the average number of cotyledons as 113 and states that they vary from 94 to 142. Craig[17] states that they are most numerous in the horns and fewer and smaller in the body of the uterus. Rainar dsaid he could only distinguish between 30 and 40 cotyledons in the uterus of heifers but after parturition 100 or more, but Williams[84, p. 135] states that this is due to their inconspicuousness in females which have not been in oestrum or pregnant and that if they were developed after birth they would tend to be irregular, but this is not so. Craig[17] states that the cotyledons are arranged in rows: 4 series in the body, 3 in the middle of the horns and 2 at the anterior extremity. This we are able to confirm.

The relation of the ovary to the opening of the Fallopian tube in the cow has been described and figured by Zucherkandl (121) and the structure of the Fallopian tubes by Schmaltz (83) who found that this varied slightly in different regions.

(c) *The Cervix.*

The cervix of the cow, which is about 4 inches long, has a very firm thick wall and the lumen is spiral in form with four annular ridges or folds (sphincters). The internal surface of the cervix is covered with lamellae or, as described by Franck-Albrecht (56), "palma plicatae" of wrinkled mucosa, the whole mucous membrane functioning as a mucous gland. A very complete description of the anatomy of the cervix, with illustrations, has been given by Williams (84, p. 28) who states that a study of the cervix of the cow is sufficient to refute allegations of the entrance of the penis into the cervical canal which is an anatomical impossibility as the channel is very small and twisted. Albrechtsen (120) also figures the appearance of the external os in the cow. Rigidity of the cervix in the cow is not a pathological but a physiological condition (Paimans (122)).

It is a very difficult and tedious operation to insert even a very thin catheter into the os of a cow; hence the difficulties of artificial insemination in the cow as compared with the mare where the os is comparatively large. Albrechtsen (120) however states that during oestrus the os is more open and a large catheter can enter it; we however have been unable to insert a probe the whole length of the cervix under these conditions but only into the external part. In order to obtain entrance the catheter has to be inserted with a rotating movement.

The changes occurring in the cervix at different stages of the oestrous cycle are shown in Table VI; it becomes congested during oestrus, *i.e.* slightly before the uterus and the congestion, which is not very marked, disappears 48 hours after the commencement of oestrus. Williams (84) states that the muscular walls of the cervix relax during oestrus and the cervical canal dilates. Weber (6) found that during heat the os was always more or less open.

One of the chief characteristics of the cervix is the mucous secretion (or slime) which it produces, the amount and consistency of the mucus varying with the stage of the oestrous cycle. The mucus produced is very thick and viscid and a relatively small amount is formed from 3 days after heat to 3 days before the next one. Just before heat it becomes fluid and large quantities are formed; during heat it is very thin and clear, becoming thicker and rather whitish, due (see p. 99) to the admixture of leucocytes and debris, 48 to 72 hours after the beginning of heat.

Emrys-Roberts (112) found that in the cow the profuse mucinous secretion during pro-oestrus is derived not from the body of the uterus but from the cervix and vagina. Krupski (91) found in cows killed at the time of heat that the os was open and contained clear fluid slime.

Bland Sutton (123) has figured and described the production of mucus by the cervical epithelium in Macaque monkeys and also in woman.

Blair Bell (124, pp. 49, 67, 70) states that in woman mucus is secreted at the beginning of the menstrual flow and he figures mucous glands in the cervix very similar to those observed in the cow. Stockard and Papanicolaou (30) found that in the guinea-pig the vagina is filled with a clear frothy mucus at the beginning of heat in addition to the vaginal plug found after coitus in these animals. Keller (125) found a mucous secretion from the cervix of the bitch at the time of heat.

The significance of these changes in the cervix during the cycle will be seen when comparison is made with that of the pregnant animal (see p. 164) in which the entire cervix is plugged with a large quantity of very thick sticky mucus. The oestrous cycle may be likened to pseudo-pregnancy or a miniature pregnancy culminating with oestrus instead of parturition, the changes occurring being similar although less in degree.

During the middle of the cycle the cervix contains thick mucus, developed probably as a result of the action of the corpus luteum, but in smaller amount (corresponding to the smaller duration of the corpus luteum) than in pregnancy. Just before heat and similarly just before parturition the cervix dilates and the mucus liquefies and streams down the vagina to the vulva. In the latter case (parturition) the passage of the cervix is dilated and lubricated for the passage of the foetus while in the former case (heat) the dilution and flow of mucus allows a free passage in the cervix and forms a liquid medium in which the spermatozoa can swim freely on their way to the Fallopian tubes.

Possibly one of the reasons why acidity of the vagina is a cause of sterility is due to the action on the consistency of the mucus of the cervix; mucus in a slightly acid medium becomes tough and stringy while in a slightly alkaline one it becomes fluid. Hence the benefits obtained by washing out with dilute alkalis before service.

Dr H. E. Woodman has confirmed by chemical tests the fact that the cervical secretion consists almost entirely of mucin and is now making a study of its properties in detail.

(d) The Vagina.

Williams (84, p. 33) states that in the cow the vagina is about 22 cm. long and that during manual exploration of the organ, more especially at the time of oestrus, it becomes ballooned owing to an inflow of air at the moment of inspiration. The details of the anatomy of the vagina have been well described by Schmaltz (83).

Normally no hymen exists, but Baumgärtner (126) has described it in a foetus of 22 cm. long and occasionally it persists in adult animals causing retention of the secretions (white heifer disease).

The vagina may be roughly divided into three areas. The portion next the os is very like the latter in the character of its epithelium (one or two cells thick) and in appearance, being very wrinkled and thrown up into ridges. The portion between the vulva and the opening of the urethra is covered with stratified epithelium very like the mucous membrane of the mouth. The portion just above the urethra combines the character of these two parts. Owing to the variation in the characters of the different parts of the vagina a separate description of each is given in Table VI which shows the changes occurring during the oestrous cycle.

In slaughtered animals the portion next the vulva nearly always shows a deep red colour some way under the surface owing to the large number of blood vessels supplying the clitoris and labiae; the surface capillaries however show some dilation just before and more especially just after the period of oestrus. The main area of congestion in the vagina however is just above the urethra. This is very slightly congested just before heat; the congestion becomes more marked during heat, is very congested 24 hours after heat and reaches its maximum engorgement 48 hours after the beginning of heat when in certain animals bleeding takes place. Similar changes but to a smaller degree take place in the upper end of the vagina, the mucous membrane being quite pale until during heat it is slightly congested and this rises to a maximum with, in many cases, bleeding 48 hours after the beginning of heat.

The vaginal secretions of all cows do not contain blood after oestrus. Of 18 heifers ranging in age from $1\frac{1}{2}$ to 3 years old 12 bled regularly, one (C 2) occasionally and 5 not at all. In the 4 cows observed no sign of bleeding was seen. This is in agreement with the opinion of herdsmen (Appendix I (3)); some say that cows in good condition bleed more than others but the majority consider that there is no difference in this respect.

Whether different breeds vary is not known but two Guernsey herdsmen have informed me that bleeding is not very frequent in their breed whereas Friesian and Devon herdsmen have told me that practically all their heifers and most cows bleed. Weber(6) found that heifers bled more frequently after heat than cows and that service has no effect on this post-oestral secretion (see p. 167); bleeding appears after the animals have been served by the bull in the same way as in those which have not been served, but like menstruation in women it ceases during pregnancy.

Krupski(91) found that out of 73 heifers killed within four days of heat 55 bled and of 79 cows killed during the same period 38 only showed signs of bleeding; the cause of the less frequent bleeding in cows he attributes to the thicker blood vessels of the uterus of animals which have been once pregnant.

In our opinion however the bulk of the blood flow comes from that part of the vagina situated just above the urethra but small amounts may be derived from other sources, uterine and cervical as well. Why bleeding should be more frequent in heifers than in cows and occur in some individuals and not in others is probably dependent on the thickness and toughness of the epithelium which increases with age, since extravasation of blood occurs in the tissues of most. Also the pendant (abdominal) position of the uterus and vagina in cows as compared with its pelvic position in heifers would tend to prevent the flow from the vulva in old animals.

The presence or absence of bleeding does not seem to be correlated with the intensity of heat (although Weber(6) and some herdsmen (Appendix I (3)) believe this is the case) for most cows which show symptoms of intense heat do not bleed whereas heifers showing only slight symptoms frequently do so.

The time after the beginning of heat that blood appears from the vulva varies in different individuals and in different cycles of the same individual as will be seen from Table XII. The average time that blood appears is 62 hours after the beginning of heat, the range of variation being from 46 up to 80 hours after the commencement of oestrus.

In those heifers which do bleed the duration of the blood flow appears to be greatest after a long cycle (see Table XII) so although whether or no an animal bleeds does not depend on the length of the cycle yet if she does bleed the intensity of the bleeding would appear to be dependent on the length of the cycle.

Table XII. *Relation of blood flow to oestrous cycle in heifers.*

No.	Details of each cycle				Average for individuals			
	Length of cycle hrs.	Duration of oestrus hrs.	Time that bleeding occurred after beginning of oestrus	Duration of bleeding	Cycle hrs.	Oestrus hrs.	Blood after oestrus	Duration of bleeding hrs.
P 1	—	—	3 days	—	—	—	3 days	—
P 2	—	—	2 days, 2 days 2 days, 2 days	—	—	—	2 ,,	—
P 3	—	—	3 days	—	—	—	3 ,,	—
C 1	—, 502 526	19, 20 18	3 days, 4 days 70 hrs.	1 day, abt. 12 hrs. 12 hrs.	514	19	80 hrs.	16
			Hours	Hours				
C 2	427	7	46	12	427	7	46 ,,	12
C 4	—, 432, 470 440, 490, 436	16, 12, 12, 10 22, 12	45, 52, 74 58, 74, 58	—, —, — —, —, —	454	14	60 ,,	—
C 5	—, 436, 428 458, 438, 442	14, 16, 16, 16 18, 30	74, 63, 58 50, 48, 62	12, —, — —, —, —	440	18	59 ,,	12
C 6	—, 424, 412 456, 454, 424	14, 14, 18, 8 22, 22	67, 66, 74 64, 72, 64	—, —, — —, —, —	434	16	68 ,,	—
C 7	—, 436, 474 508	22, 22, 18, 22	60, 55, 62 65	—, —, —	473	21	60 ,,	—
P 5	abt. 508, 544	14, 10	60, 48*	60 (18) 82* (16)	544	12	54 ,,	71 (17)
P 6	abt. 432, 448	20, 14	58, 50*	18 (14), 20* (6)	448	17	54 ,,	19 (10)
P 7	abt. 464, 464	20, 20	60, 62*	12 (12), 16* (16)	464	20	61 ,,	14 (14)
P 8	abt. 456, 502	10, 18	62, 64	6 (6), 8 (8)	502	14	63 ,,	7 (7)

* After fertile service.   ( ) Continuous bleeding.

Grouped by length of cycle.

Cycle length hrs.	No.	Average cycle length hrs.	Average duration of oestrus hrs.	Average onset of bleeding from beginning of oestrus hrs.	Average duration of bleeding hrs.
520–560	2	535	14·0	59·0	47·0
480–520	5	502	19·2	71·8	26·7
440–480	10	459	17·0	61·8	13·7
400–440	12	430	15·4	58·3	15·0

Individual averages grouped by cycle.

Longest	10	475	17·4	62·3	37
Average	6	439*	16·3	60·2	—
Shortest	8	442	12·2	47·7	28

* Some individuals not represented in average group.

From Table XII it would appear that on the average the longer the cycle the longer it takes for blood to make its appearance after the beginning of oestrus. It has been shown above that a short cycle is associated with a short oestrus, or the end point of oestrus is hastened, and correlated with this the onset of bleeding is hastened and its duration shortened. As the shortened oestrus has been shown to be associated with the abbreviated effect of the corpus luteum it is natural to conclude that the bleeding is the result of the same cause since it behaves in the same way (see p. 21). These conclusions obtain support from two facts elucidated in the rabbit: (1) that congestion of the uterus occurs at the end of pseudo-pregnancy (Hammond and Marshall(60)) and (2) that the time of abortion after removal of the ovaries in pregnant rabbits increases with the period of pregnancy (24) for as has been pointed out above it is considered that the changes occurring at oestrus, the end of pseudo-pregnancy and parturition are similar in kind but vary only in degree.

The portion of the vagina situated next the os undergoes very similar changes to those that occur in the cervix, the consistency of the mucus produced being thin during heat but becoming thick 3 days after it and remaining thick until just before the next oestrus. The mucus lying in the area differs from that of the cervix however in that it contains in addition the products of the vagina—blood and leucocytes —so that during the middle or end of heat, it contains yellowish lumps of leucocytes and 48 hours after heat streaks of blood from the vaginal walls. Later, 3 days or more after heat it becomes dark brown, the colour being due to the decomposing blood. Weber(6) states that in all except one of the cases of cows observed during oestrus the slime from the vagina was clear or serous; the one exceptional case had blood stained mucus but this animal had a vaginal cyst.

We have occasionally observed a trace of blood in the secretions during oestrus after service by the bull (see Appendix II, p. 203) but this condition is not usual.

(e) *The Mammary Glands.*

Good accounts of the structure of the udder of the cow have been given by Fürstenberg(127) and more recently by Zietzschmann(128) and Rubeli(129).

The udder of the cow, which consists of four quarters or glands, is supported by a central band of connective tissue which forms a complete barrier between the left and right sides. Bitting(130) states that each gland is enclosed in a capsule of elastic tissue which yields readily to change of form produced by the filling and emptying of the gland.

Supernumerary or rudimentary nipples and glands may exist (Schikele(131)) and many have been observed in the course of this investigation (see Plate XXVIII, 3). Bitting(130) states that these rudimentary glands, which may vary in number from 1 to 5, are generally situated posteriorly to the hind nipples, occasionally occur between the fore and hind nipples and are very rarely seen in front of the fore quarters. These supernumerary glands spoil the shape of the udder and occupy space that would be better filled by development of the normal glands. Both Henneberg(132) and Burckhard(133) found that supernumerary teats in cows were very frequent, about 38 per cent. of all cows having them. They found that the supernumerary teats may vary from 1 to 4 in number and are usually situated behind the normal nipples but may occasionally occur between the normal ones. Rather over 50 per cent. of the cows in this country have supernumerary teats(134).

Henneberg found that they occurred in certain breeds more frequently (Fleckvieh in 53 per cent.) than in others (E. Friesian and Dutch 27 per cent.) and as the latter breeds are those which have been specialised for milk production, he concluded that their presence is not associated with large milk production but that they lower the milk yield by causing the other glands to be smaller than normal.

Burckhard (133) found that the accessory teats occurred in both sexes and he has figured many cases of supernumerary nipples on the scrotum of bulls. A case of the inheritance of supernumerary nipples from a bull to his daughters has occurred at the University Farm, Cambridge, within the last few years. Bell(135) has shown that supernumerary teats are strongly inherited in sheep.

The fore- and hind-quarters on each side are quite separate as can be seen by injecting each with a different coloured gelatine solution, as has been shown by Bitting(130). Rubeli(129) seems to doubt this however, but the better results obtained by "cross milking"—*i.e.* right fore and left hind milked together—may be explained by means (see p. 63) other than connection between the fore- and hind-quarters. As a rule the hind-quarters are rather larger than the fore-quarters, as can be seen in stained sections of glands in the earlier stages of development before

Table XIII. *Amount (c.c.) of milk obtained from different quarters of the udder.*

Average of 4 days' milkings, quarters milked in different order each day.

Cow		Morning Right		Morning Left		Evening Right		Evening Left		Total Right		Total Left	
		Fore	Hind	Fore	Hind	Fore	Hind	Fore	Hind	Fore	Hind	Fore	Hind
Shorthorn:	12th week of lactation	1275	1357	1100	2082	797	862	722	1425	2072	2220	1822	3507
Jersey:	15th ,, ,,	1432	1032	712	1820	987	1120	550	1285	2420	2152	1262	3105

Percentage of total milk obtained from the udder.

	Right side Fore	Right side Hind	Left side Fore	Left side Hind	Both sides together Fore	Both sides together Hind
Shorthorn	21·6	23·1	18·9	36·4	40·5	59·5
Jersey	27·0	24·0	14·0	35·0	41·0	59·0

PLATE IV

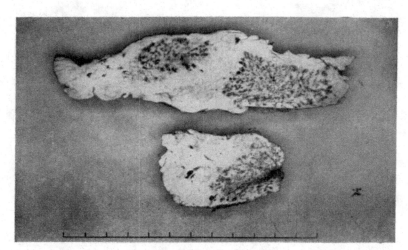

1. Udder of calf.

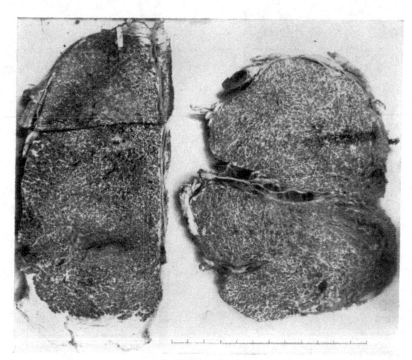

2. Udder of cow.

they have grown sufficiently to interlock; this can also be determined by the amount of milk that can be obtained from each. Table XIII shows the amounts of milk produced from different quarters of the same cow and it will be seen that the fore-quarters yield approximately 40 per cent. and the hind 60 per cent. of the milk produced; in only one case was a fore-quarter better developed than the corresponding hind-quarter. Beach and Clark[136] found that on the average of 15 cows the fore-quarters yielded 42 per cent. and the hind-quarters 58 per cent. of the total yield. Bitting[130] states that as a rule the fore-quarters end abruptly but the hind-quarters are usually prolonged upwards and backwards to a point; this we can confirm.

How far the shape of the udder as seen to external appearances affects the development of the glands is not known. No opportunity has been obtained of measuring the amount of milk produced from the fore- and hind-quarters of the typical Ayrshire udder which is long and flat and comparing it with the amounts produced from the quarters of the typical Jersey udder which is rounded. The efforts of Friesian breeders have been directed towards increasing the width of the udder rather than its length. Plumb[137] states that the yielding capacity of the udder can be increased by extending the fore-quarters but he gives no data to support this statement. Whether the prolongation of the fore-quarter in a well-shaped udder is caused by additional gland or by fat development is not known. Investigations of this point would show how far the standards of excellence of shape of the udder set up by breed societies were of economic importance. Hansen[13] and Eckles[138] have figured many udders of abnormal shapes but they give no data of the amount of milk produced as a result. Zwart[139] found that the milk pressure was capable of increasing much more in rounded udders than in flat ones.

The milk cistern is large in the cow and Wing[140] states that it holds up to half a pint of milk. Wirz[141] quotes Fleishman to the effect that the size of the cistern is 250 c.c. and he himself measured the holding capacity of the udder by injecting gelatine; one fore-quarter took 1380 c.c. and a hind-quarter 1360 c.c. the total capacity of the udder being 5510 c.c. These facts are of importance in determining whether or no milk is produced during the process of milking beyond that of the storage capacity of the udder. We have taken casts of the milk cistern in plaster of Paris and found it to be 400 c.c. in one case of fairly large size; the size of the cistern however varies greatly in different individuals and in different quarters of the same udder.

Rubeli[129] divides the cistern into two parts—the teat portion which is cylindrical and the gland portion which is variously shaped but usually short and broad; the line of junction between these two parts is constricted by a band of connective tissue and circular veins. Bitting[130] found that the cistern may be more or less divided into pockets by constrictions (see also Paulli[142]) and that the size of the milk cistern may vary greatly in different udders; he figures one in which the gland area is large but the cistern small and another in which the cisterns are large but the gland area poorly developed.

The practical importance of these facts is probably that an udder with a large gland but small duct system may suffer a greater decrease in rate of secretion after a long interval between milkings owing to the high milk pressure produced. Nevertheless Fürstenberg[127] from an examination of two cases concluded that the size and form of the milk cistern has no effect on the amount of milk secreted.

There is much room for investigation of the way in which the duct and alveolar area varies with the stage of lactation and the effect of milking at different intervals in connection with it.

Bitting[130] found that the large ducts anastomose freely but the small ones do not. Riederer[143] found that the number of large ducts entering the cistern is usually 9–11 but Zwart[139] found 15 in a Dutch cow. Whether the number of these bears any relation to the yield of an animal is not known. Wirz[141] states that the number of large ducts entering the cistern is usually 8–12 and by injecting a metal alloy into the nipple he was able to obtain casts which showed a neck-like constriction of the mouths of the ducts next the cistern and beyond this a dilation from which many secondary ducts originated, these also having constrictions at their point of entrance into the main duct. Both Bitting[130] and Wing[140] state that the branching points of all ducts both large and small are governed by sphincter muscles but that they cannot be entirely closed; by means of these the cow is able to "hold up" her milk.

Nüesch[144], Zwart[139] and Zeitzschmann[145] have put forward theories as to how the milk is held up by cows and they all agree that it is effected by erection by the vasomotor system through reflex action, but do not agree as to the exact way in which it is produced. We have examined histologically the mouths of the large ducts leading from the milk cistern and have found much smooth muscle tissue beneath the epithelium of the cistern and round the mouths of the ducts but not in the upper part of the ducts. The muscle at the mouths of the large ducts appears to surround blood vessels rather than to form a circular sphincter round the exit of the duct. These facts have a bearing on the mode of action of pituitary extract on the milk flow (see Hammond (146) and Gavin (147)).

Christ (148) describes a plentiful supply of elastic fibres round the alveoli, ducts and cistern and also describes muscle fibres as numerous round the "stricht-canal" of the teat, less plentiful round the cistern and large ducts and only occasional fibres between the alveoli. The latter we believe are in connection with blood vessels rather than the milk ducts.

Wirz (141) and others have shown that the milk ducts are surrounded by a network of elastic fibres and he believes that the collapse of the mammary gland on milking is brought about by elasticity such as occurs in the lungs and is not due to any muscular contraction. This would explain why quick milking is usually attended by a greater flow (Bitting (149) and Crowther (150)) and why the quarter milked first usually gives the largest yield (Beach and Clark (136) and Emery (151)) and the highest percentage of fat (Crowther (150)). These facts we can confirm. If elastic recoil is the origin of these variations in amount and composition as a result of quick milking it would be expected that the variations would be greatest in the earlier stages of lactation when the largest amounts of milk are produced. Sticker (152) has called attention to the elastic fibres in the walls of the milk ducts of the cow but describes no muscular fibres. It appears probable that the flow during milking is assisted by all three factors, the muscular walls of the cistern contracting, the erection of the blood vessels causing pressure on the alveoli and the elasticity of the duct walls assisting the flow from them. Nüesch (144), Zwart (139) and Zeitzschmann (145) have described the flow of milk in the cow as occurring in two parts: (1) a small continuous secretion during the interval between milkings and (2) a rapid secretion during milking, but how far this appearance is due to the anatomical structure of the gland in controlling the rate and composition of the flow has been discussed by these authors and it would appear to afford a profitable sphere of investigation from an economic point of view. Rubeli (129) states that on filling the udder with fluid (formalin or alcohol) only about 60–70 per cent. of the milk yield obtained at a milking can be forced in.

Bitting (130) states that the size of the nipple is independent of the size of the gland and that it is very elastic and provided with a copious vascular system; the nipple under excitement such as occurs in nymphomania or sometimes during milking may become turgid or erect; this is produced by the arteries becoming distended with blood. In a collapsed state the walls of the canal of the nipple lie in longitudinal folds which are obliterated when the teat is fully distended. Rubeli (127) describes the nipple between milkings as having a small cavity and engorged veins but during milking the cavity is filled with milk and the veins are comparatively small. Bitting (130) states that an averaged sized teat holds 1–1½ oz. of milk. The size of the nipple varies considerably in different individuals and breeds, being small in the Ayrshire and Jersey and large in the Red Poll and Friesian. The nipple is closed at its apex—the "stricht-canal"—by a strong sphincter and at its junction with the milk cistern also there is a constriction. Bitting (130) states that this upper constriction cannot completely close the canal and that occasionally there is a third constriction nearer the cistern; these upper constrictions are usually involved in cases of imperforate quarters and we have observed this in one case in which the nipples were injected with different coloured gelatines. Constrictions of this sort in the teat of the cow have been described and figured by Hug (153).

The structure of the nipples in the cow has been well described by Fürstenberg (127) who first figured the rosette formed by the folds of the mucosa between the "stricht-canal" or narrow opening of the teat and its hollow cavity; a detailed histological account of the "stricht-canal" has been given by Mańkowski (154).

The changes in the form of the cavity of the teat and the vascular network of the nipple during the act of milking are well illustrated by Rubeli (129) and Reiderer (143) has given a detailed account of its anatomical and histological structure.

Käppeli (155) investigated the comparative structure of the teat in many of the domestic animals and found that the muscular sphincter occurs mainly in those species with large milk cisterns.

Both Fürstenberg (127) and Riederer (143) describe the nipple as consisting of two parts, a lower portion without hair and an upper portion covered with fine hair. Fürstenberg has pointed out that only the former exists in heifers and that the latter is not developed until just before the first calf is produced; this part, the base of the teat, really consists of the lower portion of the milk cistern and in many old cows is very large and puffed.

The "milk veins" run on each side from just in front of the fore quarters along the belly to the "milk well" a hole in the abdominal wall situated near the xiphoid cartilage of the sternum; it is believed by some

PLATE V

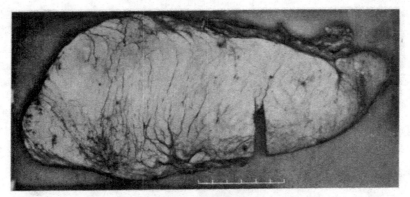

1. Udder of freemartin.

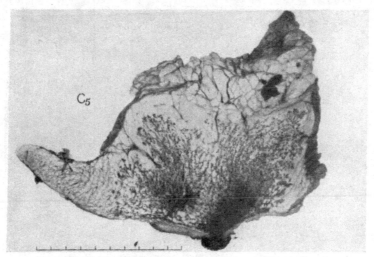

2. Udder of heifer during the oestrous cycle—3 days before heat.

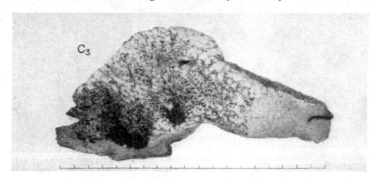

3. Udder of heifer during the oestrous cycle—about 14 hours before heat.

that their large size and sinuous appearance forms a means of detecting a good milking cow. Aldrich and Dana [156] measured the milk veins and compared the size with the yield given by the cow; they found variations in size with the stage of lactation and an increase in size and crookedness in young cows after calving; but the correlation between the size of the vein and the milk yield was not very marked. In our opinion the size of the milk vein varies with the stage of lactation, the blood pressure and the thickness of the skin of the animal and is therefore not a reliable means of judging a cow for milk production. Moreover, Fürstenberg [127] and Bitting [130] who have given a good description and figures of the circulation in the udder state that the veins anastomose so freely that blood from any part of the udder may pass back to the heart either by the milk veins (subcutaneous abdominal) or by the external pudic vein, and Bitting for this reason believes that too much value is placed upon these veins in judging an animal; one of the factors which tends to increase their size is the pregnant uterus pressing on the iliac veins. Graves [157] ligatured the milk veins (subcutaneous abdominal) of two cows to see whether the posterior veins (external pudic) would carry all the blood away from the udder. In one cow that had been in lactation 18 months he found no shrinkage of the flow after tying one vein and only a slight drop after the second was ligatured a week later. In the second cow which had calved two months and was giving 44 lbs. of milk per day the milk flow was only slightly lowered by the operation.

Nelke [158] found that the blood vessels of the udder of a good milking cow are so large that if no pressure is exerted in them they will contain from 47 to 57 per cent. of the total blood in the body and he says that milk fever is brain anaemia due to the accumulation of blood in the udder; the injection of air by pressure forces the blood from the veins of the udder. He found that although the most general time for milk fever was soon after calving cows can have it as late as 18 months after the last parturition and as early as the 6th month of pregnancy (after the alveoli are formed, see p. 173). Keim [159] also came to very similar conclusions. Seitter [160], however, basing his conclusions on blood pressure readings concluded that milk fever was due to vasomotor collapse through action of internal secretions and that air inflation assisted these indirectly.

The escutcheon, or arrangement of hairs on the perineum, since Guenon (see Dechambre [161]) put forward his theory, has been used by some as a basis of selection for milk production. While no direct evidence has been collected on this point it is difficult to see how it could affect the mammary glands, for hair whorls exhibiting well marked patterns exist also in other parts of the body, in the region of the armpits in animals and on the back of the neck in man. Bitting(130) found, and we can support his statement, that the escutcheon bears no relation to the size of the gland as it may extend several inches above or on either side of the hind glands. Bakker(162), who grouped a number of Friesian cows by their escutcheons and then compared the average yield of each group, found that the escutcheon afforded no guide as to the milk producing capacity of a cow.

The method of study used in our investigation of the mammary gland has been to weigh the udder in a fresh condition, the two halves of the udder being dissected off separately, as can easily be done by cutting along the elastic septum separating the right and left sides. Observations were made on the character of the secretion obtained from the nipple and the two halves were then fixed in 10 per cent. formalin. In one half of the fixed udder a moderately thin central and vertical longitudinal section through both nipples and glands was made with a long sharp knife. These sections were then washed, stained with Delafield's haematoxylin, again washed in distilled water, this time slightly acidulated with HCl, dehydrated in alcohol and cleared with the glycerine mixture given on p. 38. Thus a section of the whole udder was obtained cut, stained and treated in the same way as a microscopic section. The haematoxylin picked out the glandular portion of the udder leaving the remainder, the fatty and connective tissue, practically unstained so that the amount of growth and distribution of the gland could be seen at a glance (see Plates IV–VII).

The weights of the udder at different ages are given in Tables VIII, X and XI. In calves which have not reached the age of puberty (Table VIII) the weight varies with age from 220 gm. at 2 months old to 760 gm. at 6 months, averaging 540 gm. at 4 months old.

In virgin heifers which have ovulated (Table X) the weight varies more or less with age and the fatness of the animal from 1420 gm. at 2 years old to 2650 gm. at 3 years old, averaging 1930 gm. at 2 years 3 months old.

In cows which have produced a calf the weight of the gland varies greatly according to the period of lactation. Nearly all the cows killed (Table XI) were quite or practically dry and under these conditions the weight varies with the individual, age, and state of fatness of the animal, varying from 1680 gm. to 8980 gm. and averaging 5100 gm.

PLATE VI

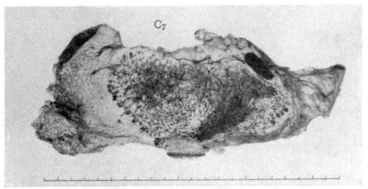

1. Udder of heifer during the oestrous cycle—6th hour of heat.

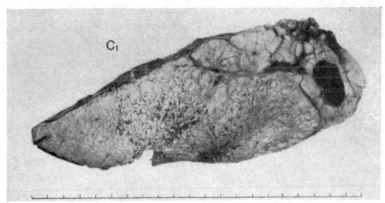

2. Udder of heifer during the oestrous cycle—24 hours after beginning of heat.

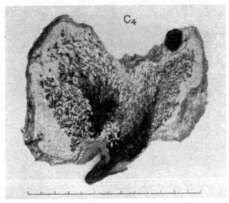

3. Udder of heifer during the oestrous cycle—48 hours after beginning of heat.

Two with diseased udders which were killed in an early stage of lactation weighed 12,000 and 12,690 gm.

The udder of one freemartin (or heifer without an ovary) was obtained, aged 3 years, in fat condition and the udder of this weighed 6400 gm.

Considering these results as a whole it will be seen that the udder, as with other parts of the body, increases in weight with age. Not sufficient material has been collected to show exactly how it varies with the stage of lactation or with the fatness of the animals. The weight and external appearance of size of an udder is however no guide as to its milk producing capacity for in the case of the freemartin quoted above, which was in fat condition, the udder weighed 6400 gm. compared with that of 2650 gm. for a normal heifer of the same age. When these two were compared by the section method (Plate V, 1 and 2) it was seen that the former contained no more mammary tissue than that of a calf of 6 months old (Plate IV, 1) the bulk of the udder consisting of fat, while the latter contained about seven times as much mammary tissue as the calf.

It will be seen (Plate IV, 2) that the udder of a cow which has produced a calf and which is just dry contains a very large proportion of glandular tissue. Therefore in order to estimate the value of the udder for milk production it is necessary to obtain in addition to the weight and size the proportion of mammary tissue present.

It is generally said that a bad fleshy udder does not shrink so much after milking as a good one which collapses to a small size after the milk is withdrawn; but how far this is due to the proportion of gland tissue, duct tissue, white fibres or elastic connective tissue is not known. Armsby[163] states that a large proportion of connective tissue causes a fleshy udder. Wing[140] states that the size of the udder in the cow is no guide to its milk producing capacity as it may be fleshy or fat and that the band of elastic tissue which separates the two sides and supports them may become relaxed so that the udder falls to the ground and may appear to be of enormous size. Bitting[130] believes however that the appearance of the udder as regards size depends on the strength of the abdominal wall which, if loose, causes the udder to fall downwards and backwards and so appear much larger. This factor he considers accounts for the sudden appearance of a large udder in some cows after the 2nd or 3rd calf, the udder becoming more pendulous as the abdominal wall relaxes. Fürstenberg[127] found that the udders of cows which show a large amount of fat tissue gave little milk and he has drawn attention to the elasticity of the skin over the udder of good milking cows; this belief is widely spread in practice. The elasticity of the skin and connective tissue probably affects the milk pressure and by relaxation reduces the pressure and so does not retard the rate of secretion. From a comparison of the amount of mammary tissue present in a calf which had not ovulated (Plate IV, 1) with a heifer which had ovulated (Plate V, 2) and a freemartin of the same age with no functional ovary (Plate V, 1) there can be no doubt that the mammary tissue develops greatly as a result of ovulation and the formation of a corpus luteum in the ovaries. This has been found to be the case in *Dasyurus* (O'Donoghue [164]) and in the rabbit (Bouin and Ancel [165], Schil [166] and Hammond and Marshall [60]). Pearl and Surface [167] found that in a cow in which the ovaries were cystic and contained no corpora lutea the udder was very small and shrunken although she had previously produced calves and milk. A very much greater growth of the gland occurs however between a virgin heifer of $2\frac{1}{2}$ years old (Plate V, 2) and a cow which has produced a calf (Plate IV, 2); the growth of the gland during this stage will be discussed below under Pregnancy (p. 168).

Since it has been found that in *Dasyurus* and the rabbit the growth of the mammary gland is correlated with the development of the corpus luteum considerable attention has been given to the changes occurring in the udders of virgin heifers at different times of the oestrous cycle. Table XIV gives the data collected on this point from heifers killed at different periods of the oestrous cycle (and see Plates V–VII). In addition observations were made during the cycle in a number of different individuals to determine the amount of fluid that could be expressed from the nipples (Table XV).

It was found that the amount of fluid that could be expressed from the nipples varied greatly in different individuals, which suggests that it may be possible to pick out good milking animals by the amount of secretion they produce as virgins at 2 years old. No opportunity however has been obtained to carry this out and it must be left for future work to decide. The individual differences being so great it was not surprising to find in the slaughtered animals differences in the amount of the secretion which could not be attributed to the period of the oestrous cycle. The observations made on living animals (Table XV) in different stages of the cycle bore out the conclusions reached from Table XIV in that the amount of the secretion varied much more in different individuals than it did during the oestrous cycle. It may be that cycles in the cow overlap so much that such marked differences

PLATE VII

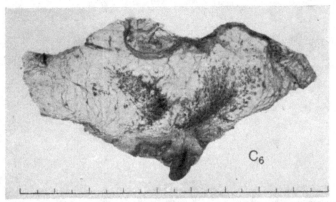

1. Udder of heifer during the oestrous cycle—72 hours after beginning of heat.

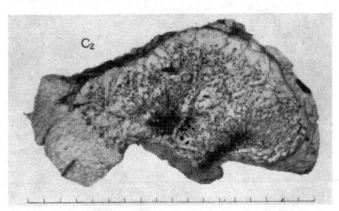

2. Udder of heifer during the oestrous cycle—8 days after heat.

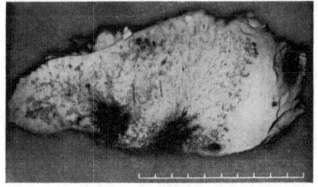

3. Udder of heifer, 1 month pregnant.

Table XIV. *Details of mammary glands of heifers killed at different periods of the oestrous cycle.*

Period of cycle when killed	3 days before heat due	14 hrs. before heat due	6th hour of heat	24 hrs. after beginning of heat	48 hrs. after beginning of heat	72 hrs. after beginning of heat	8 days after beginning of heat
No. of heifer	C 5	C 3	C 7	C 1	C 4	C 6	C 2
Age (yrs. and mths.)	1–9	2–0	2–0	2–3	2–0	2–3	2–3
Av. cycle length (hrs.)	452	455	473	502	462	445	423
Av. oestrus duration (hrs.)	18	16	21	18	13	16	8
Order of fatness: (1) Fat (7) Lean	3	1	4	7	5	6	2
Mammary gland:							
Weight of udder (gm.)	3540	1870	1720	2180	2440	1240	2740
Secretion from nipples when killed	Much in all nipples	Much in all nipples	Some fluid from all	No fluid	Much fluid from all	Little from one, none from others	Little from left hind, none from others
Section of udder — Fluid in cistern	Lot in cistern	Lot in big ducts	Little in big ducts	None visible	Very large quantity in cistern and ducts	None visible	None visible
Anatomical development of alveoli	Majority small, few swollen at tips	Small, shrivelled	Majority small, some swollen at tips	Small, shrivelled	Swollen at tips	Small and shrivelled, few swollen at tips	Small and shrivelled

in the amount of ovarian secretion do not occur as in the rabbit where long periods may elapse between ovulations (pseudo-pregnancies). The amount of the secretion obtained varied also with the time since the previous withdrawal; fluid that is produced is not altogether absorbed but remains in the milk cistern, thereby hindering further production of fluid.

The data available to support these various points is very scanty but the secretion is now being investigated in detail with accurate measurements by Dr S. A. Asdell.

Examination of the udder sections of heifers killed during the oestrous cycle (Table XIV and Plates V–VII) seemed to show that the amount of the secretion varied with the development of the milk cistern and large ducts rather than with the activity of the alveolar cells as seen in microscopic sections, and it may be that the amount of the secretion present in different individuals is a measure of the duct and milk cistern development rather than the alveolar development.

Enquiries have been made of herdsmen as to the cause of the variability in the amount of the secretion that could be obtained from virgin heifers, but their replies varied considerably (Appendix I (7)); some said that there was more in good heifers, others that it was a bad thing and led to gargetty udders later and some also thought it could be increased by feeding more especially on grass.

Table XV. *Observations made on fluid obtained from nipples of heifers in different sexual states.*

No. of heifer	Date	Sexual state		Observations on fluid from nipples
C 1	Dec. 1	—	—	Teats sore, no fluid but felt full
	„ 4	—	—	No fluid
	„ 6	On heat		„
	„ 9	—	—	„
	„ 27–30	—	—	„
	Feb. 3	On heat		— —
	„ 4	Killed		No fluid
C 2	Dec. 28	—	—	„
	„ 30	On heat		„
	Feb. 3	„		— —
	„ 11	Killed		Little from left hind, none from others
C 3	Dec. 9	—	—	No fluid
	„ 10	On heat		„
	„ 26	—	—	„
	„ 29	On heat		„
	Feb. 5	„		— —
	„ 24	Killed		Much fluid in all nipples
C 4	Apr. 29	On heat		Serous fluid from nipples
	May 17	„		Fluid from nipples
	„ 24	—	—	Little fluid
	July 14	—	—	Much fluid from nipples
	„ 16	On heat		— —
	„ 18	Killed		Much fluid from all nipples
C 5	May 22	On heat		Little fluid from nipples
	„ 24	—	—	Little fluid
	June 3	—	—	Fluid in small quantity from one nipple
	„ 8	On heat		— —
	July 18	„		— —
	Aug. 2	Killed		Much fluid from all nipples
C 6	May 23	On heat		— —
	„ 24	—	—	Much fluid from nipples
	June 3	—	—	Much from one, none from others
	„ 11	On heat		— —
	„ 29	„		Large quantity of serous fluid
	July 16	„		— —
	„ 19	Killed		Little from one nipple, none from others
C 7	May 27	On heat		Fluid from nipples
	July 27	On heat. Killed		Some fluid from all nipples
P 5	Nov. 3–6	—	—	Trace in one nipple
	„ 7	On heat		Fluid from one nipple
	„ 8–10	—	—	Little fluid from one nipple
	„ 12	—	—	Fluid from one nipple
	„ 15, 25	—	—	Little fluid from one nipple
	„ 29	On heat. Fertile service		— —
	„ 30	—	—	Much fluid from one nipple
	Dec. 1	Pregnant		Small amount from nipples
	„ 4	„		„ „
	„ 18	„ 1 month		Little fluid „ „
	Jan. 9	„ 2 months		„
	Mar. 13	„ 4 „		2 c.c. fluid
	Apr. 18	Killed 5 „		Viscid fluid from fore-quarter

Table XV (*contd.*).

No. of heifer	Date	Sexual state		Observations on fluid from nipples
P 6	Nov. 6	—	—	Much fluid from all quarters
	,, 7	—	—	Fluid from all quarters
	,, 8–9	—	—	,,
	,, 10	On heat		Little fluid from all nipples
	,, 12	—	—	Fluid from all nipples
	,, 15	—	—	Little fluid from all nipples
	,, 25	—	—	,, ,,
	,, 28	On heat. Fertile service	—	—
	Dec. 1	Pregnant		Much serous fluid from all nipples
	,, 4	,,		Moderate quantity of fluid from all nipples
	,, 12–13	,,	1 month	Fluid from nipples
	Jan. 9	,,	2 months	Much fluid from nipples
	Mar. 13	,,	4 ,,	45 c.c. fluid collected from nipples
	May 2	,,	6 ,,	100 c.c. Very viscid honey-like secretion. Left fore rather less than others and of more fluid consistency
	June 12	,,	7 ,,	All nipples full of white milk-like fluid. 280 c.c. collected from one nipple
		Killed		
P 7	Nov. 8–9	—	—	Trace from one hind quarter, none in fore
	,, 12	—	—	No fluid
	,, 15	On heat	—	—
	,, 16	—	—	Little fluid from nipples
	,, 25	—	—	No fluid
	Dec. 2	—	—	A drop of fluid
	,, 4	—	—	Little serous fluid collected
	,, 5	On heat. Fertile service	—	—
	,, 12	Pregnant		Little fluid collected
	,, 18	,,	1 month	,, ,,
	Jan. 9	,,	2 months	Some fluid collected from nipples
	Mar. 13	,,	4 ,,	8 c.c. fluid collected
	May 4	,,	6 ,,	10 c.c fluid and little left in quarters. Right hind nipple fluid secretion and other three with honey-like secretion
	,, 23	Killed	6 months	Right hind-quarter with fluid secretion, others with honey-like secretion
P 8	Nov. 11	—	—	No fluid
	,, 15	On heat	—	—
	,, 16, 25	—	—	No fluid
	Dec. 1, 4	—	—	,,
	,, 6	On heat	—	—
	,, 12, 18	—	—	No fluid

The chemical composition of a secretion obtained from a virgin goat has been found by Hill(168) to be similar to that of normal milk, but in this animal very large amounts of the secretion were produced. Woodman and Hammond(169) have investigated the composition of the secretion obtained from the above mentioned heifers and found that it contained the characteristic proteins of colostrum (globulin and albumen) together with slight amounts of fat, casein, lactose and proteose. Thus the initiation of mammary gland activity in the cow is not necessarily dependent on pregnancy but may be associated with the occurrence of the oestrous cycle. This supports Heape's(170) contention that the source of the stimulus which causes the development of the mammary glands and the secretion of milk is dependent on the ovaries and not on the foetus entirely. Numerous writers have described milk secretion in various species before pregnancy; Bitting(130) in the cow, Lowenthal(171) in the mare, Oliver(172) and Lindig(173) in women, Hill(174) in the goat and Barnowsky(175) and Leblanc(176) in various species. Sheild(177) has quoted a number of cases in girls where precocious mammary development was associated with the appearance of the menstrual flow; he states that in women after the menopause the breasts usually shrivel, in some women however they increase in size but this is due to a subcutaneous deposition of fat and the true mammary tissue atrophies.

That the secretion depends on the ovaries rather than the uterus has been shown by Foges(178) and by Hammond and Marshall(60). Moreover Steinach(179) and Athias(180) found that ovaries transplanted into male guinea-pigs caused the mammary glands to develop; the latter attributes the growth of the mammary gland to the ripe or atrophic follicle. Myers(181) has shown that in rats the development of the mammary glands at the time of puberty can be prevented by starvation and Jackson and Stewart(182) found in animals so treated the ovaries fail to develop normally.

Whether or no the proportion of weight of the ovaries (in the maximum stage of the oestrous cycle) to weight of the body varies in good and bad milkers is not known but it would not appear to be improbable.

5. The Histological Changes.

Introduction.

The histological changes were determined by cutting sections of the material which had in most cases been fixed in formalin (in a few cases formol picro acetic was used) and embedded in paraffin wax. Delafield's or iron haematoxylin and eosin were the stains most frequently used, but for the preparations in which the distribution of mucin was determined iron haematoxylin and mucicarmin was used.

A bird's-eye view of the various histological changes occurring in different stages of the oestrous cycle is given in Table XVI.

The histological appearance of the various parts of the reproductive organs of cows has been described by Schmaltz(83) to which account

Table XVI. *Histological appearance of reproductive organs of heifers killed at different periods of the oestrous cycle.*

			3 days before heat due	14 hrs. before heat due	6th hour of heat	24 hrs. after beginning of heat	48 hrs. after beginning of heat	72 hrs. after beginning of heat	8 days after beginning of heat
Stage of cycle:									
No. of heifer:			C 5	C 3	C 7	C 1	C 4	C 6	C 2
Hrs. after beginning of heat that blood appeared from vulva in previous heat periods			Bled 59, 48 and 74 hrs.	Did not bleed	Bled 55, 65 hrs.	Bled 80 hrs.	Bled 60, 45, 74 hrs.	Bled 68, 64, 72 hrs.	Bled occasionally 46 hrs.
Organ	Part	Tissue							
Ovary	Largest follicle	Granulosa	Cells small	More swollen	Swollen in thin layer, inclined to split	Swollen	Thin layer of compressed cells	Swollen. Thick layer	Deep layer
		Blood vessels of theca	Visible at base of granulosa. Compressed	Occasional at base of granulosa. Compressed	Capillaries moderately large. Large b.v. compressed laterally	Capillaries dilated. Large b.v. Swollen but compressed laterally	Not obvious	Very occasional	Large and rounded
	Most recent corpus luteum	Luteal cells	Large, cytoplasm about 5 times nucleus	Smaller and lighter staining (haem. and eos.)	Large, cells shrunken and clear. Majority squashed together, cytoplasm about 1½ times nucleus	Majority small, few large and clear	Small, densely packed and darkly staining	Cells larger but still staining darkly. Patches of colloid between cell masses	Very large and more lightly staining. Cytoplasm about 10 times nucleus
		Blood vessels	Several capillaries	Few capillaries	Few capillaries. Large b.v. more prominent and with thick walls	Few capillaries. Large b.v. with very thick connective tissue walls	Red corpuscles and leucocytes free in tissue. No formed blood vessels	Capillaries formed and numerous	Capillaries numerous
		Connective tissue	Marked layer between luteal cells	Double layer between luteal cells. Thickened strands in places	Much between luteal cells. Large strands in places	Large proportion between luteal cells. Large strands in places	Ingrowing strands from theca. None between more central lutea cells	Few strands	Single layer between luteal cells

Ovulation

Table XVI (contd.).

Stage of cycle:			3 days before heat due	14 hrs. before heat due	6th hour of heat	24 hrs. after beginning of heat	48 hrs. after beginning of heat	72 hrs. after beginning of heat	8 days after beginning of heat
No. of heifer:			C 5	C 3	C 7	C 1	C 4	C 6	C 2
Organ		**Part**							
Ovary	Previously formed corpus luteum	Tissue							
		Luteal cells	Very few left. Small	—	—	—	Majority very small and shrivelled. Few large clear cells	—	—
		Blood vessels	Mostly large b.v. with very thick connective tissue walls	—	—	—	Capillaries few, large b.v. prominent and walls thickened with connective tissue	—	—
		Connective tissue	Very large proportion	—	—	—	Many large strands and much between cells	—	—
		Connective tissue stroma	Dense	Dense	Not quite so dense	Not quite so dense	Not so dense	Not so dense	Fairly dense
		Capillaries	Few	Few	Slightly more	Not quite so marked as C 7	Great congestion and extravasation under epithelium	Great congestion and extravasation but not so much as C 4	Several capillaries, but not congested
		Pigment	Some	—	—	—	—	Traces	Much
Uterus	Cotyledon	Subepithelial connective tissue	Dense	Dense	Dense	Fairly dense	Looser	Loose	Moderately loose
		Subepithelial capillaries	Few	Few	Few	Few	Congested	Large quantity of extravasated blood in subepithelial connective tissue and here and there blobs of it under epithelium	Few
	Intercotyledonous area	Subepithelial pigment	Traces	Traces	—	Present	—	—	Little
		Deep connective tissue	Moderately loose with little hyaline material in places	Loose with much hyaline material in places	Loose with fair amount of hyaline material	Fairly loose, with some hyaline deposit being absorbed	Moderately dense with slight traces of hyaline deposit	Moderately dense. No hyaline deposit	Moderately dense. No hyaline deposit
		Relative sizes of glands — Diam. lumen	10·0	11·2	7·0	5·0	4·7	6·0	17·2
		Relative sizes of glands — Height of epithelium	5·5	4·5	5·5	6·0	6·7	5·2	7·2

Ovulation (between C 1 and C 4).

Organ	Location	Structure					Ovulation		
Cervix		Epithelium	Cells cubical.	Cells cubical.	Cells little more than cubical.	Cells columnar.	Cells columnar (long).	Cells little more than cubical.	Cells cubical.
		Mucus	Much external mucus, little in cells	Thin layer on outer side of cells. Outer edge of cell wavy	Thin layer on outer edge of cell. Wavy free streams	Much in end of glands, small quantity free at base of dips	Much mucus in ends of cells. None free	Moderate quantity in cells. Much free mucus	Little in cells and much free mucus
Vagina	Next os	Blood vessels	Normal	Normal	Large ones dilated, small ones visible	Large b.v. dilated	Normal	Normal	Normal
		Epithelium	Normal	Normal	Cells swollen	Cells swollen at base of dips	Cells filled with mucus	Normal except at base of dips swollen	Normal. External mucus in some places
	Above urethra	Blood vessels	Normal	Large vessels congested, small b.v. just visible	Extravasation of blood under epithelium. Few leucocytes	Little blood free under epithelium. Pigment under epithelium	Some blood still free, remainder being collected into capillaries	Small capillaries visible under epithelium	Large extravasation of blood under epithelium
		Epithelium	Normal	Normal	Swollen in places	Cells very swollen	Few swollen cells	Normal, growing at base	Normal
	Next vulva	Blood vessels	Large vessels congested	Large and small vessels congested	Large and small vessels congested. Leucocytes under the epithelium	Extravasation of blood	Blood collected into capillaries	Small vessels just visible	Small vessels just visible
		Epithelium	Normal	Normal	Cells swollen	Cells very swollen	Cells flaking off	Lower layers growing	Normal
Mammary gland	Alveoli	Lumen of alveoli	Filled with secretion	Majority filled with secretion	Large lumen	Moderately large lumen, alveoli small	Smaller lumen, alveoli small	Smaller lumen	Quite small lumen
		Epithelial cells	Cubical	Cubical	Cubical surface uneven	Cubical	More than cubical	Almost columnar	Columnar

the reader is referred for a detailed systematic description and references to the older literature on the subject.

(a) *The Ovaries*.

A description of the histological appearance of the ovary in the cow has been given by Schmaltz (83) and Williams (84, p. 40). Three areas in the stroma can be observed; the tunica albuginea which lies just under the germinal epithelium and is usually free from follicles; the parenchymatous layer which occupies the greater part of the ovary and in which most of the follicles lie; and a very vascular layer usually free from follicles along the area of its attachment to the broad ligament. The main histological interest centres round three points, namely (1) the ripening of the Graafian follicle, (2) the formation and structural changes undergone by the corpus luteum and (3) the origin and development of the interstitial cells.

The ripening of the Graafian follicle. The Graafian follicle of the cow is peculiar in that the theca interna is very well marked and it becomes more distinct and dense when the follicle is increasing in size.

It has been held that the theca is a specialised part of the ovarian stroma and there is no doubt that it supplies through its well developed capillaries and lamellar cells the nourishment required by the developing follicle. This well marked dense appearance is due however not only to specialisation of the connective tissue elements but to pressure caused by the increase in size of the follicle, for the theca becomes more marked as the follicle begins to increase in size. Also the area next the follicle— the theca interna—is much more dense than that on the outer side—the theca externa—between which and the ovarian stroma the limits are very indistinct; this is the natural result of pressure by the developing follicle on the ovarian stroma. The results of pressure can be seen if the connective tissue of the theca interna is observed in different stages of the ripening of the follicle. Some 18 days before ovulation (Plate XXI, 4) the connective tissue surrounding the follicle consists of a dense mass of cells with nuclei close together whereas just before ovulation (Plate IX, 1) the connective tissue in this area is drawn out and fibrous and the nuclei are a long way apart.

It may be contended that in this area there are a larger number of theca luteal cells than in other parts of the ovarian stroma, but with pressure the number in a given area would naturally be increased and the presence of lipoid granules in their interior is the natural course of the flow of this material to supply the developing ovum with yolk. The "theca luteal cells" are only modified lamellar cells of areolar connective tissue. Wolz (183) concluded that the function of the theca interna was the nutrition of the granulosa and for this purpose it contained reserve supplies of nutrition. Zschokke (40) found in the cow that the theca interna is much thicker in the large follicles than in the small ones.

Another well marked change which occurs in the theca interna during the cycle is in the number and size of the capillaries. Some 18 days before ovulation they are hardly observable (Plate XXI, 4) and the larger blood vessels in this area are rounded; about 3 days before heat the capillaries can be seen here and there in the theca and the large blood vessels are flattened by pressure; at the time of heat the capillaries become larger and more numerous and by 24 hours after heat (Plate IX, 1) they are very large and congested and lie close beneath the granulosa cells, the larger blood vessels are however at this time still flattened by pressure although greatly swollen.

Since ovulation occurs soon after this and the newly formed corpus luteum has a large central blood clot which could only be derived from these capillaries it would appear that their congestion with consequent oedema and softening of the surrounding tissues plays a large part in the sudden increase in size of the follicle at this time (see p. 41) leading to turgidity and rupture. That actual rupture of these blood vessels and extravasation of blood into the cavity is the cause of rupture of the follicle is doubtful for, as is shown below (p. 183) the thecal blood vessels may rupture and bleeding may take place into the follicle without ovulation occurring. This happens frequently in the rabbit, which does not ovulate spontaneously (24).

The bleeding into the cavity of the follicle therefore, although it may help rupture, is not the cause of it. The dilation of the blood vessels probably however assists in the formation of the secondary liquor folliculi which increases the turgidity of the follicle before rupture (Robinson (51)). Whether the turgidity in connection with a specialised point of dehiscence of the follicle (Joss (84)) or a digestive ferment appearing in the follicle (Schochet (184)) is the cause of rupture we have no evidence to offer; but it is difficult to imagine the ferment acting only on the outer side of the follicle without also digesting the granulosa cells. In addition ovulation in women is now generally believed to occur not during menstruation but about 10 days after its cessation.

Simon (109) from an examination of the ovaries of cows concluded that the follicle normally ruptures by infiltration of lymph from the blood vessels of the theca, the cells of the limiting membrane of the follicle swelling; this conclusion was based on comparison of normal follicles with follicular cysts in which the thecal blood vessels had almost disappeared and the connective tissue hardened. Käppeli (105) who made a large number of measurements of the sizes of the follicles in cows' ovaries concluded that ovulation is caused by a large accumulation of liquor folliculi following vasomotor nerve stimulus as a result of sexual excitement during heat. Delestre (100) found that in the cow as the time of rupture of the follicle approaches a collagenous layer appears between the granulosa and theca interna. Zietzschmann (26) from histological examination of cows' ovaries concluded that the hyperaemia of the genitals during heat assists the rupture of the follicle by the addition of fluids causing swelling and also by causing cell proliferation and enlargement in the granulosa and theca interna.

Long and Evans (32) in the rat found that before the follicle ruptures there is inpushing of the follicle wall by blood vessels and then a final swelling of the follicle occurs.

Stockard and Papanicolaou (38) concluded that in the guinea-pig rupture of the follicle occurs as a result of congestion which begins in the theca at the same time as in the other reproductive organs. Loeb (43) found that in the guinea-pig the corpus luteum prevents ovulation although it does not hinder the ripening of the follicle. The presence of true interstitial cells in the ovary has been pointed out by Bouin and Ancel (185) to be associated with ovulation only after coitus. Since these cells are supposed to have an internal secretion somewhat similar to that of the corpus luteum it may be that the additional congestion of the genitals following coitus is necessary in the case of species with well developed interstitial cells, this taking the place of normal congestion following degeneration of the corpus luteum in animals with no interstitial tissue. In this connection the findings of Corner (107) in pigs and Long and Evans (32) in rats that the rupture of all follicles takes place simultaneously is suggestive. However on this hypothesis it is difficult to account for the first ovulation that occurs at puberty or ovulation in monoestrous animals.

The corpus luteum. Much discussion has taken place as to the origin of the luteal cells. Whether they are formed from the granulosa or from the theca interna. In the freshly formed corpus luteum of the cow 48 hours after heat commenced (Plates IX, 2) and XI, 1 the central cavity was filled by a blood clot surrounding which on all sides except the point of rupture were luteal cells very small, densely packed and deeply staining, all of approximately the same small size and in appearance like swollen

PLATE VIII

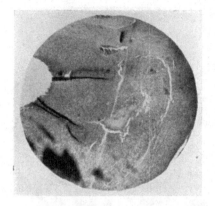

1. Corpus luteum: 72 hours.   2. Corpus luteum: 8 days.

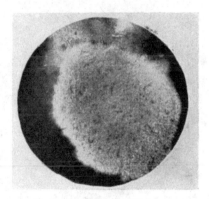

3. Corpus luteum: 21 days.   4. Corpus luteum: 23 days.

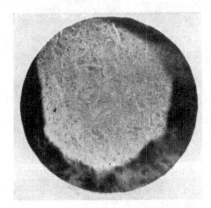

5. Corpus luteum: 37 days.   6. Corpus luteum: pregnant 8 months.

granulosa cells. Between the masses of these cells and the blood clot could be seen streams of albuminous fluid in which red and white corpuscles were freely distributed. The theca interna appeared to have completely vanished as would naturally be expected owing to relaxation of pressure brought about by rupture of the follicle, for it has been shown above (Table VII) that the corpus luteum at this stage has only about half the diameter of the follicle from which it arises. The theca interna is however to be seen in the form of ingrowing strands of connective tissue which fold in along the ovarian edges of the collapsed follicle. This stage of the corpus luteum in the cow has been well described by Zietzschmann (26) who has figured it showing the extravasation of blood in the tissues at the point of rupture, the formation of luteal cells from the granulosa and ingrowths or folds of connective tissue from the theca. He also figures the following stage with a central coagulum and formation of capillaries between the luteal cells.

In the next stage of our series—72 hours after the beginning of heat (Plates VIII, 1 and IX, 3)—the luteal cells have become larger and are not quite all the same size but are still darkly staining. Organised blood capillaries are now being formed between the masses of luteal cells and their development appears to be associated with the absorption of the pockets of albuminous fluid which have collected in the centre of the new corpus luteum. Connective tissue threads are now beginning to grow from the thick strands of the former theca interna in between some of the individual luteal cells.

By 8 days after the commencement of heat (Plates VIII, 2 and IX, 4), when the corpus luteum has reached its maximum development, the majority of the luteal cells have become very large (the cytoplasm of the cell being about 10 times the area of the nucleus) and only an occasional small one here and there is seen. The luteal cells are also not so deeply staining (haematoxylin and eosin) as before, this being due to the larger quantities of lipoid granules which they contain. They are now divided into clumps of 3 or 4 cells surrounded by a single layer of young connective tissue cells, among which blood capillaries are very numerous.

Three days before the next heat period (or 17 days after the last) the appearance is very similar to the last stage but the connective tissue has now grown in between each individual luteal cell. The luteal cells are all fairly large, the cytoplasm of the cell at this time being about five times the area of the nucleus.

About 14 hours before and also during heat (Plates VIII, 3 and X, 2) the majority of the large luteal cells have become smaller and shrivelled, those that remain large being clear or lightly staining and filled with lipoid; in the shrunken cells at this time the cytoplasm of the cells is only about 1½ times the area of the nucleus. A thick layer of connective tissue now separates the luteal cells and very few blood capillaries are visible. The larger blood vessels which have hitherto escaped notice owing to the large number of luteal cells between them now appear prominent as a result of the shrivelling of the luteal cells; moreover their walls have become greatly thickened by connective tissue growths, similar to those described below in the cotyledons of the uterus (see p. 92).

In old corpora lutea 23 days old (or 2 days after the next heat period) (Plates VIII, 4 and X, 2) only a very few large luteal cells are present, the majority being shrivelled so that the connective tissue between them frequently appears in thick patches which completely surround the blood vessels, the latter now being cut off from direct communication with the majority of the luteal cells.

Similar but more marked changes were observed in a corpus luteum 37 days old (Plates VIII, 5 and X, 3) (or 17 days after the next heat period). In this case only a few luteal cells remained, the bulk of the corpus luteum consisting of connective tissue; the large blood vessels which remained and were prominent had very thickened walls and their lumen was reduced by connective tissue growth.

Kaltenegger[186] found similarly in the cow that the connective tissue increases with the age of the corpus luteum and crowds out the luteal cells. The description given above of the changes occurring in the corpus luteum of the cow is on the whole similar to those found by Corner[107] in the sow. Schmid[16], who examined the corpus luteum of the cow at different stages of the cycle, concluded that atrophy commenced through death of the capillary network. Van der Stricht[187] has described both the serous and lipoid secretions formed by the corpus luteum in the bat.

Vignes[188] believes that the active internal secretion produced by the ovary is adsorbed by the lipoids (lecithin) and is freed from this combination and rendered active by cholesterine.

The significance of the histological changes occurring in the life of the corpus luteum in relation to its internal secretion, as determined by the effect it produces on the other reproductive organs (uterus and mammary gland), appears to us to be as follows:—The granulosa cells which have hitherto been (1) producing the secretion of liquor folliculi and (2) supplying lipoids to the ovum to build up the yolk, now for the first time come into direct contact with the blood supply from the ruptured capillaries of the theca; they increase greatly, first in number

PLATE IX

1. Wall of large follicle.

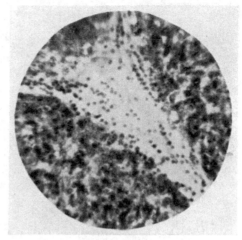

2. Corpus luteum: 48 hours.

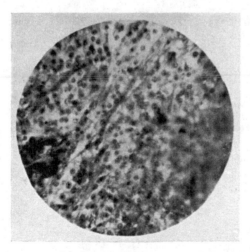

3. Corpus luteum: 72 hours.

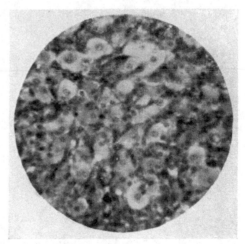

4. Corpus luteum: 8 days.

and then in size. The products of their metabolism (1) the liquor folliculi accumulates at first in the central cavity of the corpus luteum until such a time as the capillaries have become organised sufficiently to carry it away and to supply the body with this probably active hormone. Their other product, (2) the lipoid granules, is no longer required by the ovum as yolk and is not easily transported by the blood and so is stored up in the cells (whose metabolism is increased owing to better blood supply) giving them their typical colour and appearance and adding largely to their increase in size.

During this time the connective tissue elements have been increasing greatly and either this increase cuts off direct communication between the capillaries and the luteal cells and leads to the atrophy of the latter or more likely the lipoids accumulate in the luteal cell to such an extent that metabolism is no longer possible for them and they die and are gradually absorbed by the surrounding connective tissue (for duration in pregnancy see p. 121).

The time of their death and gradual absorption occurs a few days before the next heat period when the control of the corpus luteum over the developing follicle and other reproductive organs begins to disappear.

If the above theory of the internal secretions of the corpus luteum is accepted it might explain why the injection and administration of corpus luteum preparations has generally failed to give results (Corner and Hurni (46)) in any way similar to those obtained by the formation of this gland in the body (Loeb (47)), the lipoid which is stored being inert in its action and the liquor folliculi which is not stored being the active principle. It also explains why a well developed follicular system in the ovary often causes similar but less marked effects on the uterus than the corpus luteum (Zietzschmann (26) and Nielsen (90)).

Many of the experiments which have been made on the injection or administration of luteal, ovarian, foetal or placental extract are therefore open to the criticism that they were performed on animals with functioning ovaries and that variations in the activity of the latter may have affected the results obtained, cf. Marshall and Jolly (189). Wintz (190) obtained congestion of the vulva in rabbits after long continued injections with corpus luteum extract but not with follicular extract whereas Sonnenberg (191) found congestion of the genitals about 20 minutes after injections of liquor folliculi into rabbits. Schickele (192) found that extracts of the ovary and uterus of woman and the cow contained antithrombin and substances which reduced blood pressure. These were present before puberty but not after the menopause when no follicular

H   6 system was present. The antithrombin was abundant in the maternal blood of the placenta in pregnancy. Solorieff (193) found that subcutaneous injections of ovarian extract produced a secretion of colostrum in guinea-pigs but he could not obtain it after corpus luteum injections.

The effect of the corpus luteum in causing follicular atrophy considered in the light of the above explanation of the processes concerned would appear not to be a direct internal secretion effect or a mechanical effect on the follicle but to be due to the fact that both follicle and corpus luteum utilise the same source of nourishment, of which only small quantities are available in the body. In times of lack of nutrition (Loeb (57), and Papanicolaou and Stockard (31)) or of excessive utilisation of the necessary substance (as occurs in lactation and pregnancy) the amount present is not sufficient to support large follicles and they atrophy; the demands of the more primitive tissue—young follicles and corpora lutea—are satisfied first, maintenance coming before production.

That the regression and death of the luteal cells is due to the accumulation of lipoids receives support from the following facts: (1) the shortening of the cycle in fat animals and (2) the concentration of the pigment with the life of the corpus luteum. It also might explain why very fat animals do not often come in season (Wallace (194)) and are frequently sterile, the ovarian cells becoming clogged with these waste products (Marshall and Peel (53)) and so failing to function properly. Animals which are not bred from regularly would also tend to accumulate these products in their ovaries and consequently be less fertile than those which bred regularly. The accumulation of waste products in the ovary may also explain why a rising condition of nutrition is generally attended by greater fertility than a stationary or lowering state of nutrition from fat to lean condition. Again the general breeding seasons of animals are usually times of good nutrition following bad— spring after the cold winter and autumn after the hot summer—and under bad conditions fat disappears from the body.

That the fatty substances which accumulate in the corpus luteum are very similar in composition to those which form the yolk of the egg is seen by comparing the analysis of the yolk of the hen's egg (none being available for mammalian ova) which contains fat (25·25 per cent.) phosphorised fat (11·15 per cent.) and cholesterin (1·75 per cent.) with that of the corpus luteum of the cow which contains fat (2·99 per cent.) phosphorised fat (14·87 per cent.) and cholesterin (1·17 per cent.); the composition of both these substances is quite distinct from that of body fat (Cramer (195)). Robinson (51) found in the ferret that as the pre-

PLATE X

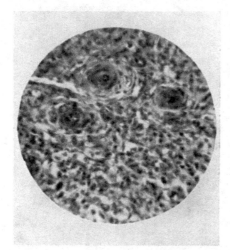

1. Corpus luteum: 21 days.

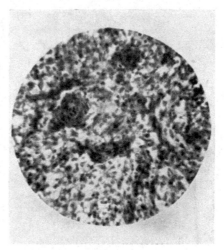

2. Corpus luteum: 23 days.

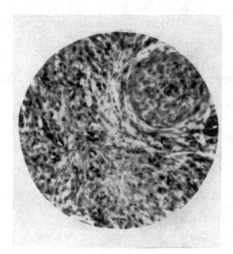

3. Corpus luteum: 37 days.

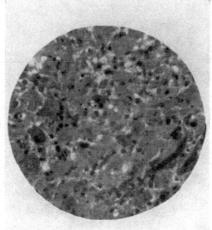

4. Corpus luteum: pregnant 8 months.

inseminal maturity of the follicle is approached granules of fat appear in the borders of the granulosa cells. Regaud and Policard (196) found in the bitch that granules were formed by the granulosa cells and were passed through the zona pellucida and accumulated in the ovum. Van Beek (197) by staining with Sudan III found lipoid globules not only in the yolk but also in the granulosa and theca interna cells of developing follicles in calves' ovaries.

As to the much debated subject of the origin of the corpus luteum we are in agreement with the theory postulated by Marshall (54) in the sheep—a related animal—that the luteal cells are formed from the granulosa cells and that the theca interna supplies the connective tissue elements and blood vessels. Zschokke (40) believed that the corpus luteum in the cow was formed from the theca interna, basing his opinion mainly on the thickness which the theca attains in the large follicle; he concluded that the reason why no corpus luteum was formed when a follicular cyst was ruptured was that the theca had lost the power of growth. In the cow Delestre (100) considers that the corpus luteum is formed from the theca entirely, but he did not examine any very early stages of the corpus luteum and based his conclusions on the appearance of the theca in mature follicles and the connective tissue central plug which forms on absorption of the albuminous secretion. Kaltenegger (186) found by the use of Unna-Pappenheim's stain that lamellar cells of the connective tissue were present in the corpus luteum of the cow. These cells, which have frequently been described as "theca lutein cells" have, we believe, often misled people into the assumption that the luteal cells developed from the theca interna and their presence in large numbers in the theca interna has given rise to the idea that it was a special tissue of internal secretion rather than one specialised for the nutrition of the follicle by concentration of certain elements of the connective tissue.

The interstitial cells. No true interstitial cells comparable to those that occur in the rabbit (which are similar in appearance to the luteal cells and are derived from the germinal epithelium (Lane-Claypon (198))) have been found in the ovaries of either calves, heifers or cows. In cows or heifers which have ovulated there are many cells in the ovary derived from old corpora lutea which have the appearance of interstitial cells, but if careful observation is made it is seen that they are in connection with a plug of connective tissue from an old corpus luteum.

Cells are frequently found in the ovaries of calves, heifers and cows which are often mistaken for true interstitial cells; they are the connective tissue lamellar cells which in the ovary frequently have granules

6–2

of lipoid deposited in them in much the same way as the lamellar cells of the frog's skin contain pigment granules or those of the silky fowl contain melanin granules. These cells are present in all ovaries and have frequently been described as interstitial cells or special "theca lutein cells" because in certain areas they have lipoid granules deposited in them on their way to the ovum or corpus luteum. Leucocytes have also sometimes been observed carrying these granules.

It is therefore concluded that the cow has no true interstitial cells comparable with those of the rabbit and derived directly from the germinal epithelium but that after ovulation a certain number are accumulated from luteal cells which have escaped degenerative changes or from atrophic follicles. Numerous lamellar connective tissue cells however exist and they frequently have lipoid deposited in them temporarily and have often been mistaken for true interstitial cells.

Aschner(199) who studied the distribution of the interstitial cells by staining with Sudan III found that in calves and heifers before puberty there were no interstitial cells and that only a few occurred in the ovaries of old cows.

Pearl and Surface(167) have described and figured interstitial cells in the ovaries of cows which had produced calves. They state that they are usually rounded in outline but may be considerably elongated in one direction, approaching spindle shape; the nuclei are of the same order of magnitude as those of surrounding connective tissue cells and they contain granules of fat stained by Sudan III and others stained by kresylechtviolett differentiated by washing in alcohol. This description appears to us to be not dissimilar to that of lamellar connective tissue cells.

They found these cells in the ovaries of a cow which had assumed secondary male characters and which contained cysts but no corpora lutea and concluded that secondary sexual characters were therefore due to the corpus luteum and not to the interstitial cells. Van der Stricht(187) states that in the bat there are no interstitial cells in the ovary of the embryo but that these appear a few days after birth; that atrophic follicles go to form the interstitial gland and the substance secreted by the interstitial gland is the same as that secreted by the corpus luteum (lipoid). In the rabbit the appearance and staining of the interstitial cells is practically identical with the cells of the corpus luteum. In the bat Athias(200) found that interstitial cells may be developed from the theca interna of atrophic follicles and also from interfollicular stroma cells. Long and Evans(32) in the rat found that interstitial tissue was developed from atrophic follicles. Popoff (201) states that in woman interstitial cells may be formed from atrophic corpora lutea but that their origin varies in different species. Loeb (43) states that in the ovary of the guinea-pig there does not exist a structure deserving the name interstitial gland; only small shrinking connective tissue cells of theca interna filling the place of lost parts of the follicles which are in process of atresia.

We are in agreement with Bouin and Ancel's (185) finding that in animals with spontaneous ovulation (including the cow) there is no true interstitial gland but that its place is taken by the periodic corpus luteum and also believe that many of the interstitial cells described are hypertrophied lamellar cells of the connective tissue such as those figured and described by Kaltenegger (186) in the corpus luteum of the cow. The presence of a lipoid secretion in them does not denote an internal secretory function but is nutritive, designed to supply yolk to the egg. No serous or liquor folliculi secretions have been observed from these cells.

Sections of the ovaries of old cows when compared with those of heifers show a much greater proportion of ovarian stroma to germinal tissue. It is considered that this has been brought about by the persistence of connective tissue developed from successive corpora lutea (see pp. 47 and 80) in addition to the normal increase in the proportion of connective tissue with age in all parts of the body. Käppeli (105) found a very much smaller number of follicles in the ovaries of cows than of heifers and still less than in calves. But decrease in the number of follicles is not the cause of the larger proportion of connective tissue for it has been shown (see p. 47) that the ovaries of cows are very much heavier than those of heifers or calves in spite of the decrease in the number of follicles present.

(b) *The Origin and Distribution of Mucin.*

One of the most distinctive features of the cycle in the cow is the copious flow of mucus from the vulva during the period of oestrus. With the naked eye it was possible to trace the main source of this to the cervix and upper portion of the vagina. In order to make quite certain however, microscopic examination of the epithelial cells in each area of the tract was made by staining for mucin.

The best results were obtained by fixing the material in formol-picro-acetic which hardens the mucin; sections were cut embedded in paraffin and were stained with iron haematoxylin and Mayer's mucicarmin (carmine 1 gm., aluminium chloride 0·5 gm., water 2 c.c. and 50 per cent. alcohol 100 c.c.) which picks out the mucin very distinctly.

The uterine glands and mucosa contribute but a small amount of very watery secretion to the flow as could be determined by observation with the naked eye. By histological examination a thin film of mucus only could be detected over the upper portion of the glands and uterine epithelium (Plate XI, 2).

The cervix, as could be seen by the naked eye, produces the bulk of the mucous secretion. Sections showed that at some stages of the oestrous cycle very large amounts of mucin were stored up in the tops of the epithelial cells (Plate XI, 3) and during pregnancy the mucus could be seen distending the spaces between the epithelial lamellae and folds of the cervix (Plate XIV, 4).

That part of the vagina situated next the os appeared to the naked eye to be the seat of a considerable amount of mucous secretion, but whether actually formed in the region or flowing over the surface from the cervix could only be determined by histological examination. The epithelium which in this area is only slightly stratified (usually one or two cells thick) is thrown up into folds and ridges and in the former the epithelial cells are generally only one layer thick. This single layer of cells contains large quantities of mucin (Plate XI, 4) but it was not present in such large quantities as in the cervix.

In that portion of the vagina next above the urethral opening the epithelium is more thickly stratified and there are far fewer folds. These folds unlike those of the previous region usually consist of stratified epithelium. Mucus was only secreted in small amounts from this area, the greater part coming from isolated pockets at the base of the existing folds (Plate XI, 5).

The vagina next the vulva consists of very deeply stratified epithelium and only a few small depressions are seen on its surface. Between the cells of the epithelium in these depressed areas a few threads of mucus could be seen, which however would only be sufficient to keep the surface moist and would not add much to the oestrous flow (Plate XI, 6).

A small amount of mucus-like secretion is produced by the cells of Gärtner's canals (Plate XXVII, 3) which lie just beneath the epithelium of the vagina and open on each side about the level of the urethral opening (see p. 97). The normal small size of these glands however precludes the possibility of their adding but a very small amount to the flow.

It will thus be seen that the bulk of the mucous flow originates in the cervix and this is supplemented by mucus which is produced in the

PLATE XI

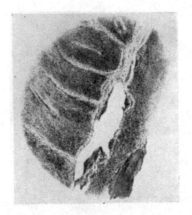

1. Corpus luteum: 48 hours.

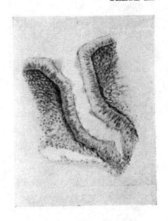

2. Uterus.

3. Cervix.

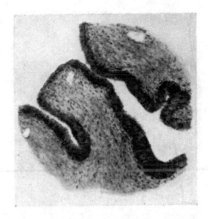

4. Vagina next os.

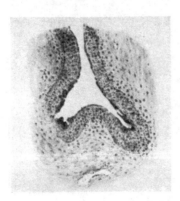

5. Vagina above urethra.

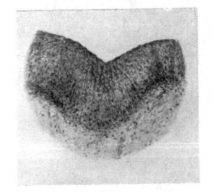

6. Vagina next vulva.

folds of the upper end of the vagina; the rest of the reproductive tract produces not much more than is sufficient to lubricate its own surface.

By the use of mucicarmin Barrington (202) found in guinea-pigs and other animals great changes in the mucin content of the vaginal epithelium in different sexual states, and increase more especially in pregnancy.

Böhme (203), also by the use of mucicarmin, found that the whole epithelium of the female reproductive organs contained some mucin, but that most occurred where the epithelium was of columnar type; he found very thick layers in the cervix and also that the amount appeared to vary with the functional state.

Moreaux (204), by means of various stains for mucin, has shown that a cycle of secretion exists in the epithelium of the Fallopian tubes of mammals.

The cyclic changes in the mucin content of the epithelium in the different parts of the reproductive tract in cows is described below (p. 92) where evidence is adduced that its production follows the cyclic changes in the ovary and corpus luteum. The enormous development of mucin in the cervix during pregnancy is described on pp. 116–164.

(c) *The Uterus.*

The mucosa of the uterus of the cow is not homogeneous like that of the majority of other animals but, in common with other ruminants, is differentiated into two parts: (1) The intercotyledonary area or general surface which is similar to the uterine surface of most animals and on to which the mouths of the glands open. Just under the epithelium there is a thin rather dense layer of connective tissue and beneath this a deep layer of more open connective tissue in which the glands ramify. (2) The cotyledons, 80 or more in number, are raised button-like projections arranged more or less in rows parallel to the longitudinal axis of the uterus; their function is to form the main points of attachment for the foetal membranes. The cotyledons have no uterine glands opening on to their surface; they are covered by a single layer of epithelium beneath which is a mass of dense connective tissue in which blood vessels are very numerous. This layer which constitutes the bulk of the cotyledon is probably a thickening of the dense subepithelial layer which is found in the intercotyledonary parts of the mucosa (Hilty (205)); beneath this layer there is another composed of open connective tissue and glands such as exists in other parts of the uterus (Wall (206)).

It is probable that as suggested by Wall(206) the cotyledon and the thick subepithelial layer of the intercotyledonary area contain connective tissue cells of embryonic type which are easily stimulated to proliferation.

Wall states that the crypts of the cotyledons are to be seen as formed structures in the virginal uterus but with this statement we do not agree.

The uterus has two muscular coats between which there is a vascular layer of loose connective tissue in which many large blood vessels ramify.

The histological changes occurring in the uterus at different stages of the oestrous cycle are shown in Table XVI and will be discussed below.

The Cotyledons. Three days and also 14 hours before heat the connective tissue stroma of the cotyledons appears to be dense and the blood vessels small (Plate XII, 1). During and 24 hours after the beginning of heat the subepithelial blood vessels in the cotyledons appear to be slightly congested and the connective tissue is not so dense. By 48 hours after heat (Plate XII, 2) there is great congestion of the blood vessels in this area and extravasation of blood into the tissues just under the epithelium; the connective tissue also appears to have become oedematous and is less dense. Seventy-two hours after the beginning of heat there is still great congestion and extravasation but not so much as before; most of it has now probably been extruded into the cavity of the uterus or is in the process of absorption. Traces of amorphous brownish pigment can now be seen, these probably being derived from the decomposing red blood corpuscles; the connective tissue is still oedematous. By 8 days after the commencement of oestrus (Plate XII, 3) the congestion has subsided although several capillaries can be seen; the connective tissue has now become more dense and the oedema has subsided. Many pigment granules, probably derived from extravasated blood, can be seen in the connective tissue; this is similar to, but not so pronounced as, the pigmentation which follows heat in the sheep (Marshall(54)).

The relation of these changes to the fixation of the foetal membranes will be dealt with under "pregnancy" below (p. 148) but it may be said here that the congestion and bleeding which occurs in the cotyledons is not a preparation for the attachment of the foetal membranes for these do not become fixed to the cotyledons for about a month, after the time when the next heat period would have been due. It is probable that the extravasation of blood into the tissues is similar to that which occurs at the end of pseudo-pregnancy in the rabbit, but in the cow the cycles (or pseudo-pregnancies) overlap.

The less dense character of the connective tissue of the cotyledon

PLATE XII

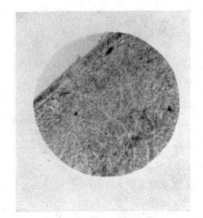

1. Cotyledon: 3 days before heat.

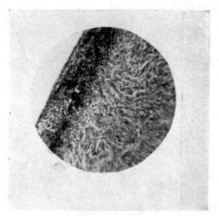

2. Cotyledon: 48 hours after beginning of heat.

3. Cotyledon: 8 days after heat.

4. Thick-walled blood vessels in cotyledon of a cow.

5. Intercotyledonary area of calf.

6. Intercotyledonary area: 14 hours before heat.

just after heat is probably due to the oedema following congestion of the blood vessels.

The Intercotyledonary Area. The changes in connection with the subepithelial blood vessels and connective tissue which occur on the general surface of the uterus are very similar but less severe than those which take place in the cotyledons. The capillaries are normal in size 3 days before heat, become slightly enlarged during heat but do not appear to reach their maximum congestion with extravasation of blood until rather later (72 hours after heat) than the cotyledons; it is possible however that this may be due to the individuality of the animals examined. Extravasation of blood under the epithelium is followed by pigment formation (8 days after heat) but not to the same extent as in the cotyledons. Kolster[207] has described leucocytosis and diapedesis of red blood corpuscles from the capillaries in the uterine mucosa of a heifer killed after heat and also the presence of pigment probably derived from blood.

In the stroma forming the deeper layers of the uterine mucosa there appears to be an accumulation of hyaline matter between the strands of loose connective tissue about 3 days to 14 hours before heat (Plate XII, 6). Twenty-four hours after the commencement of oestrus this appears to decrease in amount and to disappear 48 to 72 hours after the beginning of heat when the connective tissue becomes more closely packed (Plate XIII, 2); 8 days after heat no hyaline material is apparent and the stroma appears fairly dense.

The accumulation of hyaline material between the strands of connective tissue in association with the swelling of the connective tissue cells would appear to be the result of nutritive control by the corpus luteum, for similar accumulation of hyaline material and swelling of connective tissue cells with consequent rarefication of their nuclei occurs during pregnancy (see p. 144). This swelling of the connective tissue cells is seen in pseudo-pregnancy under the action of the corpus luteum in the rabbit, more especially after stimulation by foreign bodies, when they form decidual cells and hyaline matter accumulates in large quantities (Hammond[208]). Wells[209] states that mucoid degeneration of connective tissue seems to be merely a reversion to the foetal type; apparently when connective tissue reverts to an embryonal type, either from intrinsic causes (tumour formation) or when the nourishment is insufficient, or possibly when the normal stimulus to cell growth is absent (myxedema), the mucoid characteristics of foetal tissue reappear.

The size and appearance of the uterine glands also varies with the period of the cycle although no great difference could be observed in their number and distribution. The size of the glands and more particularly the lumen are greatest near the epithelium and become smaller in the deeper regions next the muscular coat.

Seventy-two hours after the beginning of heat the glands are rather small and their central lumen is almost obliterated; the distance between the glands appears to be large. Eight days after the commencement of oestrus the glands appeared to be much larger, the epithelium being deeper and the lumen much greater (Plate XIII, 3). Three days and also 14 hours before the next heat period the glands were smaller and the majority of them, even the deep seated ones, had flattened epithelial cells and a large lumen in which the products of secretion could be seen. During heat and more markedly 24 and 48 hours (Plate XIII, 1) after the beginning of heat the central lumen appeared to have become smaller especially in the deeper glands. At this stage the gland size is at a minimum but the epithelium is comparatively high.

The cyclic changes appear most marked in the deep glands and the size of these is greatest during the middle of the cycle and then just before the onset of heat; this is followed by a rapid diminution in size to a minimum 24–48 hours after heat at a time when congestion of the blood vessels takes place; thereafter from 3 days after heat until beyond the middle of the next period a gradual increase in size again occurs.

Since increase in size of the uterine glands has been shown to be associated with the development of the corpus luteum in pseudo-pregnant rabbits by Ancel and Bouin [210] and by Hammond and Marshall [60] and in the bitch by Keller [211] and Marshall and Halnan [73] it is believed that the corpus luteum is also the cause of the cyclic changes in size of the glands and that the cycle in the cow is comparable to pseudo-pregnancy in rabbits. The glandular hypertrophy that occurs in the uterus of heifers during the cycle is similar but less marked than that which occurs with the more long continued development of the corpus luteum in pregnancy (see p. 141).

If the changes—development of glands, blood vessels and hyaline material in the stroma—described above during the cycle are compared with those which occur during pregnancy (p. 138) it will be seen that the cycle is essentially a pregnancy on a small scale, due we believe to the extent to which the internal secretions of the corpus luteum act. The periodic corpus luteum has only a transient action on the uterus because of its short duration whereas the corpus luteum of pregnancy has a more marked effect because of its long continued and steady

PLATE XIII

1. Intercotyledonary area: 48 hours
after beginning of heat.

2. Intercotyledonary area: 72 hours
after beginning of heat.

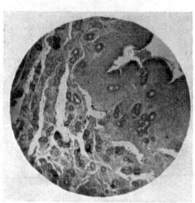

3. Intercotyledonary area: 8 days
after beginning of heat.

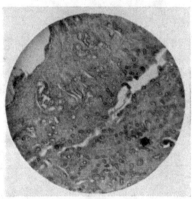

4. Intercotyledonary area: pregnant
1 month

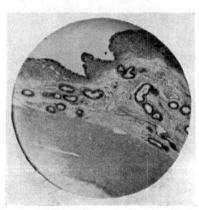

5. Intercotyledonary area: pregnant
4 months

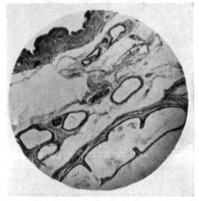

6. Intercotyledonary area:
pregnant 7 months

action. Moreover the termination of the action of the corpus luteum in the cycle which results in the extravasation of blood into the tissues and the decrease in size of the glands is similar to, but on a smaller scale than, the changes which occur at parturition.

Zietzschmann[26] states that in the cow some 2–3 days before heat the uterine mucosa begins to swell; during heat the swelling reaches its maximum and in heifers extravasation of blood into the tissues occurs after the high point of heat has been reached; blood accumulates in the subepithelial mucosa and raises the epithelium while here and there it penetrates to the uterine cavity but is mainly absorbed *in situ* forming pigment. He states that the growth of the glands ceases after heat and they begin to atrophy; growth and hypertrophy of the mucosa is taking place on the 12th day after heat and on the 13th the glands begin to secrete and the epithelium to become lower and lower, a large lumen being formed. He concludes that from two days before heat until the 12th day after it the uterus is in a state of proliferation and that from the 13th to 19th day after heat it is in a state of degeneration.

We do not, however, agree with this entirely but consider that hypertrophy begins with the development of the corpus luteum 3 days after heat and that the swelling and oedema of the mucosa which occurs about the time of heat is not a proliferative change, but is the beginning of degeneration such as occurs under pathological conditions in association with congestion. This swelling should therefore not be confused with the true hypertrophy which occurs as the direct result of corpus luteum action beginning about 3 days after heat. As pointed out above however the end of pseudo-pregnancy in the cow coincides with the pro-oestrus period.

We also believe that the proliferation continues after the 13th day, for the secretion and formation of a large lumen by the uterine glands is not a degenerative change but occurs during pregnancy when still further gland development occurs.

Hilty[205] has described the histological changes occurring during the involution of the uterus in the cow; he found that the lumen of the glands diminished rapidly from 144μ at the surface of the mucosa and 50μ in the deep glands 3 days post-partum to 45μ and 14μ respectively 21 days post-partum. The epithelium of the glands was reduced only slightly from 20μ at the surface and 16μ in the deep glands 3 days post-partum to 16μ and 15μ respectively 21 days post-partum. We consider this involution to be associated with the disappearance of the corpus luteum of pregnancy.

The uteri of calves before puberty are distinguished by the few glands which they contain and also by their less prominent cotyledons and their lack of vascularity (Plate XII, 5). The small number and size of the uterine glands of the calf as compared with the cow has been observed by Böhme (203) and is shown in figures by Kolster (207) who states that after puberty the glands grow down to the muscular coat.

The appearance of the uteri of old cows which have previously borne calves varies with the stage after parturition but is distinguished from that of heifers by the thicker muscular coats and larger blood vessels. At the base of the cotyledons and in the stratum vasculosum of the muscle especially, the blood vessels have very thick walls due to the occlusion of their lumen after parturition by connective tissue growths (Plate XII, 4). During involution of the cotyledons after parturition these blood vessels do not decrease in size so much as the rest of the stroma and consequently their proportions are increased in cows which have produced a calf.

Hilty (205) has described and figured the histological changes occurring in the post-partum uterus of the cow; he found that the walls of the blood vessels thicken and that this proceeds together with the obliteration of their lumen by growth of the connective tissue, of which many examples are figured. Three days post-partum the lumen of the cotyledonary arteries measured 140μ and their wall 160μ whereas 43 days post-partum the lumen measured only $80\,\mu$ but the wall $240\,\mu$. He also found that pigment is formed from blood which is extravasated in the cotyledon at parturition.

(d) The Cervix.

Baumgärtner (126) has shown that the cervix is well differentiated from the rest of the genital tract in the cow foetus of 22 cm. in length.

Histological examination of the cervix shows that it is composed of a mucosa thrown up into a large number of folds very similar in appearance to those of the Fallopian tubes (Plate XIV, 3). These folds or laminae have a central core of connective tissue supplied with blood vessels and are lined by a single layer of epithelium which forms simple sacculated glands. The character of the epithelium varies considerably in different stages of the oestrous cycle (Table XVI).

During the greater part of the cycle from about 3 days after the commencement of heat until just before the next heat period the cells are cubical in shape. During heat the cells became slightly longer and by 24 hours after heat are columnar and their length is still further

PLATE XIV

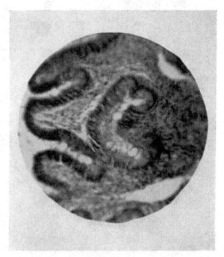

1. Cervical epithelium: 8 days after heat.   2. Cervical epithelium: on heat.

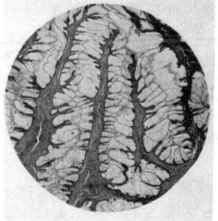

3. Cervix, shortly after heat.   4. Cervix: about $3\frac{1}{2}$ months pregnant.

increased by 48 hours after, but by 3 days after the commencement of heat they have almost returned to their cubical form.

The reason for these changes in shape can be seen when preparations have been stained for mucin. The increase in size that takes place in the length of the cell 24 hours after the beginning of heat is due to the accumulation of mucus in the outer edges of the cells, at this time only a little thick mucus being seen free at the base of the folds. Forty-eight hours after heat has commenced there is a tremendous accumulation of mucus at the ends of the epithelial cells (Plate XI, 3) but very little is visible outside the cells in the spaces between the lamellae. Seventy-two hours after the beginning of heat however secretion of mucus has taken place and the cells have become smaller but still contain considerable quantities in their outer edges; at this time the spaces between the lamellae are gradually filling with mucus discharged from the cells. By 8 days after heat has commenced the cells are again cubical with only a little mucus on their outer edge but with much free thick mucus filling the spaces between the lamellae (Plate XIV, 1). The same condition but with much more free mucus is observed 3 days before the next heat period (and in the pregnant animal tremendous accumulation of free mucus is seen (p. 165)). In the heifers killed 14 hours before and at the 6th hour of heat, at a time when the natural flow of mucus from the vulva was at its height, there was only a little mucus between the lamellae and this was not in the form of a solid mass but was loose and frothy; the outlines of the cells, more particularly their outer borders, were wavy and indistinct and contained a moderate quantity of mucus (Plate XIV, 2).

It would appear that the large amount of mucin appearing within the cell and increasing its length, starting about the time of heat until 72 hours after it has commenced, is associated with the cessation of its secretion and that this occurs at a time when the action of the corpus luteum on the generative tract is at a minimum. Beginning with the new corpus luteum 3 days after the commencement of heat and culminating with its atrophy about 3 days before the next heat period a secretion of thick mucus takes place into the spaces between the lamellae, this being intensified during pregnancy. As a result of the atrophy of the corpus luteum and the onset of another oestrus there is a general oedema of the generative organs which in the cervix is characterised by an alteration in the nature of the consistency of the mucous flow as well as by slight congestion (see p. 54) and exudation of leucocytes, thus giving rise to the oestral flow of mucus. Similar liquefaction of the mucous plug in the cervix occurs just before parturition following oedema of the organs at this time.

The cervix of calves 6 months old before the age of puberty contains a little mucous secretion between the lamellae and the epithelium is usually flattened to cubical in shape.

The histology of the cervix and its secretions in the cow and other animals has received very little attention. Blair Bell[124] figures the histological appearance of the cervix in woman and it seems to be similar to that of the cow with a mucous secretion.

Trautmann[212] has described the anatomy and histology of the cervix in the sow, mare and ewe and figures the histological appearance of the cervix in the pregnant ewe in which a large accumulation of mucus can be seen; this we have confirmed by our own observations.

In the fowl large quantities of albumen are secreted from the upper portion of the oviduct next the funnel (Kaupp[95]) but it is unlikely that this is homologous with the cervix. Mucin is present in the egg of the fowl in small quantities, but the portion of the tract from which it is formed has not yet been determined; if it is confined to any particular region it might help to explain the relationship of the different parts of the tract in birds and mammals.

The accumulation of thick mucus which takes place in the cervix and its liquefaction before oestrus must play an important part in allowing the ascent of the spermatozoa in animals like the cow, sow and ewe when the cervix is long and narrow (see Sterility, p. 180).

(e) *The Vagina.*

Three distinct areas in the vagina have been distinguished: (1) next the os, (2) above the urethra and (3) next the vulva. There is however no hard and fast line between them but the character gradually changes from vulva to os, the thickness of the epithelium decreasing as the os is approached. Wall[206] found that the spread of the stratified epithelium increases with age; in a 3 months foetus it extended only to the vestibule and the vaginal epithelium was columnar, later however the stratified epithelium grows in and at 2 years old extends almost to the cervix. In old cows he states it may extend even into the cervix. The area next the os is characterised mainly by its secretion of mucus; that above the urethra by being the seat of maximum congestion and bleeding; and that next the vulva by the thickness of its epithelium and patches of lymphatic tissue situated under it. The changes occurring during the oestrous cycle in these areas are shown in Table XVI and are described below.

(1) *The Vagina next the Os.* As has been pointed out above, in discussing the distribution of mucin, the stratified epithelium in this area is thin and the whole surface is thrown up into folds at the bases of which mucin is produced frequently by a single layer of epithelial cells (Plate XI, 4). The shape of the cells of this single layer whether normal or swollen with mucin varies at different stages of the cycle in the same way as those of the cervix. The changes in the blood vessels in this area are much the same but less intense than those of the area next described.

(2) *The Vagina above the Urethra.* This region has far fewer folds in its epithelium than the last area and with this a reduced number of mucus secreting cells. In this region the folds decrease in size until they become merely pockets in the wall; in these the epithelium is stratified and mucus accumulates in the cell at the same period of the cycle as it does in the cells of the cervix. Here however the accumulation of mucus is not followed by an increase in the height of the cell but by a swollen and globular appearance causing the distance between the nuclei to be much greater than at other times in the cycle (compare Plate XV, 1 with 3 both from the vagina next the vulva).

The changes in the vascularity of this area at different stages of the cycle are shown in Table XVI and it will be noticed that congestion begins here before it does in the uterus, the large blood vessels becoming slightly congested just before heat begins. During heat there is extravasation of blood from the capillaries situated immediately under the epithelium and many leucocytes can be seen in the connective tissue; the same appearance is shown by the heifer killed 24 hours after the beginning of heat but in addition pigment masses were present. Forty-eight hours after the beginning of heat blood could be seen on the vaginal surface and also between the epithelial cells, leucocytes were also very numerous in the epithelium and connective tissue under it. Seventy-two hours after the commencement of oestrus the extravasated blood has nearly all been set free on the surface of the vagina but a few red blood corpuscles are still seen between the epithelial cells; the majority however are being collected into capillaries again, but the leucocytosis of the subepithelial connective tissue still persists.

In basing conclusions upon the appearance of sections from animals killed just after the heat period the variation in time between the beginning of heat and the appearance of blood from the vulva in different animals has to be taken into account; this may vary from 45–80 hours after the beginning of heat (see Table XII) and sometimes may not occur at all. The heifer which was killed 8 days after the commencement of heat (C 2) was one which bled only occasionally and it may be that the large amount of extravasated blood found under the epithelium in certain places is not a normal occurrence in animals which bleed freely, although she had shown some bleeding 2 days after the beginning of the last heat.

The occurrence of subepithelial congestion in the vagina during heat probably accounts for the fact that bleeding may sometimes occur after service or when the hand is inserted in the vagina at this time, although the bleeding does not occur normally until later if the epithelium is uninjured.

In all the above heifers care had been taken not to handle the generative tract either per vaginam or per rectum during any of the cyclic changes.

It is believed that the thickening of the epidermis of the vagina of old cows is one of the reasons why bleeding does not occur in them so frequently as in young heifers.

Wall (206) states that with age in cows the stratified epithelium of the tract spreads at the expense of the columnar epithelium and quotes Sobotta who says that in woman after several parturitions the stratified epithelium is increased owing to its greater regenerative power.

It should be pointed out that in woman the menstrual flow is derived mainly from the uterus and that no great diminution is experienced with age whereas in the cow the bulk of the flow comes from the vagina and decreases markedly with age.

(3) *The Vagina next the Vulva.* The stratified epithelium in this area is much thicker than in the rest of the vagina and contains only a few small pits on its surface; these undergo in their mucin content the same cycle of changes as, although less marked than, those described in the foregoing area. The chief characteristic of the region is the large number of small lymphatic nodules which exist scattered under the epithelial surface; the enlargement of these is one of the symptoms of granular vaginitis or bacterial infection of these parts (see Hutyra and Marek (213) and Raebiger (214)). Lymphatic patches occur also in other parts of the vagina but they become less frequent as the os is approached.

The vascular changes are similar to those described above for the area above the urethra but the large blood vessels are always numerous and marked in the connective tissue below the epithelium, for they are closely connected with the vascular supply to the erectile organs of the clitoris and labiae.

PLATE XV

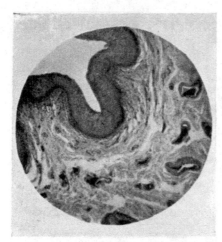

1. Vagina next vulva: 3 days
before heat.

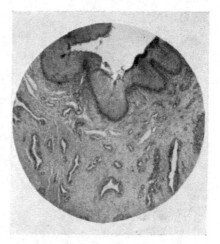

2. Vagina next vulva: 6th
hour of heat.

3. Vagina next vulva: 24 hours after
beginning of heat

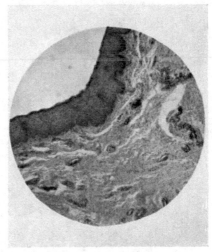

4. Vagina next vulva: 48 hours after
beginning of heat.

Three days before heat the large blood vessels are the only ones noticeable (Plate XV, 1). Fourteen hours before heat they become congested and the small capillaries lying just under the epithelium also show signs of congestion. During heat in addition to the congestion of the capillaries there is a large accumulation of leucocytes free in the connective tissue just beneath the epithelium (Plate XV, 2). No doubt the penetration of these through the epithelium is the cause of the whitish colour of the mucous flow towards the end of heat (see p. 103).

Twenty-four hours after heat there is a large extravasation of blood beneath the whole surface of the epithelium which it penetrates in places (Plate XV, 3).

Forty-eight hours after the beginning of heat, at a time when the blood flow usually commences, the leucocytes and red blood corpuscles can be seen in the epithelium itself between the cells (Plate XV, 4). Seventy-two hours after the commencement of heat, after the blood flow has taken place, the blood vessels are in a collapsed state and at 8 days after heat the tissues have assumed their normal condition.

Few descriptions have been given of cyclic changes in the vagina. Retterer [215], who examined the vaginal epithelium in cows, found that heat produces no marked changes in the cells but that changes do occur in the last half of pregnancy when the cells become cylindrical owing to accumulation of mucus. Latase [216] in rodents found a definite rhythm, alternating between epidermal and mucus, in the vaginal mucosa corresponding to the rhythm of the ovaries.

Stockard and Papanicolaou [38] have described leucocytosis and shedding of the epithelial cells in the vagina of the guinea-pig towards the end of the heat period. Long and Evans [37] have described a cycle of changes in the vagina of the rat in which a layer of epithelial cells is thrown off at each heat period. Their figures at the beginning of stage 1 of heat show the superficial cells to be large with clear vacuoles and these are greatly increased during pregnancy. Although they do not say so, it appears to us that these vacuoles are accumulations of mucus such as occurs in the cow. Although in the cow we can observe no great shedding of the vaginal epithelium after heat such as occurs in the rat yet the depth of the stratified epithelium increases about the time of heat owing to the oedema and accumulation of mucus within the cells.

Gärtner's Canals. These have been figured and described by Schmaltz [83] in the cow; they open one on each side of the vagina at the level of the urethral opening and their ducts may be seen in sections of the vagina cut above this level.

The canals lie buried in the connective tissue some little way beneath the epithelium and consist of a central slit-like duct with a few tubular glands opening into it. In those animals which have never been pregnant the canals are small and the tubular glands are in a relatively undeveloped condition (Plate XXVII, 3). The canals are lined by a thin stratified or double layer of epithelium of which the outer is often cubical in form; the glands are lined by a single layer of cubical epithelium and contain a little colloid secretion (the appearance is very similar to the thyroid).

Meyer[217] who examined Gärtner's canals in women states that they are probably rudiments of the male tract (Wolffian tract) and Baumgärtner[126] figures them as such in cow foetuses of 13 and 22 cm. It is stated by Ballantyne that the tubules of the canals are better developed in women who have borne children than in others.

During pregnancy in heifers there is apparently a large development of the gland tubules (see p. 168) and occasionally they are the source of cysts when their ducts have become occluded. This has been pointed out by Ruder[218] who figures Gärtner's canal in a calf at the level of the cervix. Dohrn[219] and Kocks[220] have also given accounts of the appearance of the canals in women and animals.

Roeder[221] examined the Gärtner's canals of a number of cows varying in age from 2 months to 16 years old and found that at any one age the length of the canals varied greatly and the canals in each side were very often unequal in length; he found that in old age they become much smaller. Cysts were found frequently which contained a yellowish clear fluid mucous secretion.

We have seen one case of a vaginal cyst caused by a distention of Gärtner's canals with fluid (Plate XXVII, 1); this was from Cow 28 which was sterile and had also follicular cysts in the ovaries. Hess[41] states that he has frequently met with them but they are not important as causes of sterility.

Barrington[222] has shown that in cats the epithelium of Bartholin's glands becomes rich in mucus shortly before heat and also during the last half of pregnancy. Bartholin's glands occur in the cow under the epithelium of the lips of the vulva but the changes occurring in them have not been determined. Marshall[223] states that these glands produce a viscid secretion during sexual excitement and Hess[41] frequently found cystic Bartholin's glands in nymphomaniac cows. The structure of the clitoris has been described by Koch[224] in the calf.

(f) The Vaginal Discharges.

The character of the vaginal discharges has been studied in detail for it was thought that it might be possible to diagnose from these the stage of the oestrous cycle and so obtain an estimate of the time a cow was due to come on heat. Stockard and Papanicolaou[38] have found that it is possible to do this in the guinea-pig and Long and Evans[37] have shown that the stage of the cycle could be determined from the vaginal discharges in the rat.

A study of these discharges might also provide a basis for the diagnosis of abnormalities of the reproductive tract caused by disease.

With the naked eye it can be seen that the character of the discharge varies at different stages of the cycle. The vagina is comparatively dry during the greater part of the cycle; a day or two before heat it becomes rather moist; at the beginning of heat it shows a flow of clear fluid mucus; and towards the middle or end of heat the flow contains small cheesy yellowish white lumps. About the end of heat the mucus becomes thicker and just after heat frequently whitish in colour and very much less in amount. In most heifers this is followed about the second or third day after the beginning of oestrus by a flow of mucus stained with blood which usually lasts from 6 to 20 hours but may continue intermittently for 3 days (see Table XII). For one or two days after the blood flow has ceased the mucous secretion is small in amount and of a yellowish white colour and then usually about 6 days after the beginning of heat the vagina appears comparatively dry and normal again, remaining in this condition until just before the next heat period.

This general description of the character of the flow will apply in most cases but time variations occur in the same way as has been shown above for the length of the cycle and for the duration of heat. Also local conditions, such as a wound made by animals horning one another in the neighbourhood of the vulva, will alter the character of the flow just as mastitis or inflammation of the udder alters the cell content of milk. Table XVII shows the notes made on the nature of the secretions during five successive cycles in the same heifer.

Table XVIII shows the character of the vaginal secretions together with the results of histological examination of smears from three different heifers at various periods during the oestrous cycle. Many cycles have been observed in this way and the following is a general account of the changes which usually occur. The method used was to take swabs from that part of the vagina next the vulva by holding apart the labiae with

Table XVII. *Character of the vaginal discharges of a heifer (C 1)*
during five successive cycles.

	Cycle				
Period of cycle	From Sept. 10 to Sept. 30	From Sept. 30 to Oct. 20	From Oct. 20 to Nov. 10	From Nov. 10 to Nov. 30	From Nov. 30 to Dec. 21
5 days before heat	Little mucus	—	—	—	Dry
4 ,, ,,	Dry	—	—	—	—
3 ,, ,,	,,	—	—	—	—
2 ,, ,,	—	—	Moist, little mucus	Fluid mucus	Dry
1 ,, ,,	—	—	String of mucus	Moist	Fluid mucus
24 hrs. ,,	—	—	—	,,	Moist
6 ,, ,,	String of mucus	—	—	—	—
Beginning of heat	Little mucus	String of mucus	String of mucus	Slight mucous flow	Moist
2nd hour of heat	—	—	—	—	Fluid mucus
4th ,,	—	—	—	String of fluid mucus	Fluid mucus with cheesy lumps
End of heat	—	—	—	Thick lumps of mucus	Little mucus
2 hrs. after end of heat	—	—	Moist	—	—
6 ,, ,,	—	—	—	String of mucus with yellow patches	—
12 ,, ,,	—	—	—	Yellow mucous discharge	Thick mucus
24 ,, ,,	—	—	—	—	,,
36 ,, ,,	—	—	—	—	Moist
2 days after beginning of heat	—	—	—	—	Blood tinged mucus
3 ,, ,,	Blood on tail	—	—	Blood and yellow mucus	Little mucus
4 ,, ,,	—	—	Blood on tail	Blood and mucus	Thick mucus
5 ,, ,,	—	—	Blood on tail and flank	—	Slight secretion
7 ,, ,,	—	—	—	—	Dry
11 ,, ,,	—	—	—	—	,,
15 ,, ,,	—	—	—	—	,,

one hand and with the other gently rubbing the surface with a cotton-wool swab on the end of a glass rod.

Films were then made from this swab and stained in various ways, but usually with Delafield's and iron haematoxylin and eosin.

A film taken about 3 days before an animal was due to come on heat, when the vagina was comparatively dry, showed only fairly numerous epitheloid cells from the vagina (Plate XVI, 1).

A few hours before heat began, and when the vagina appeared moist, the films showed an occasional leucocyte and rather fewer epitheloid cells due to their dilution with mucus.

At the beginning of heat, when there was a flow of clear mucus, films again showed only a few vaginal epithelial cells and leucocytes (Plate XVI, 2); but later in heat, when the cheesy lumps were excreted,

PLATE XVI

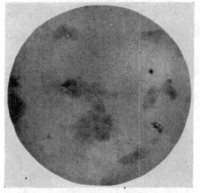

1. Three days before heat (vaginal epithelial cells).

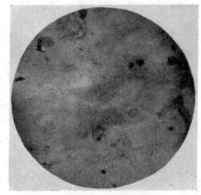

2. Beginning of heat (mucus).

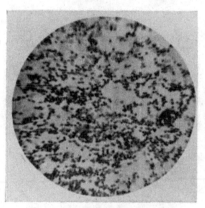

3. Seven hours after end of heat (many leucocytes).

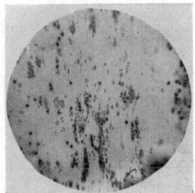

4. Seventy-two hours after beginning of heat (blood).

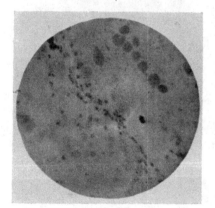

5. Thirty-four hours after end of heat (epithelial cells and several leucocytes).

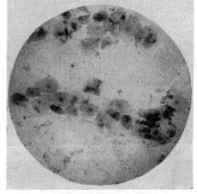

6. Fifteen days after heat (epithelial cells)

Table XVIII. *Character and microscopical appearance of the vaginal discharges of three heifers during different periods of the oestrous cycle.*

No. of heifer:	C 1				C 2				C 3			
Period of cycle	Character of discharge	Vaginal epithelial cells	Leucocytes	Red blood corpuscles	Character of discharge	Vaginal epithelial cells	Leucocytes	Red blood corpuscles	Character of discharge	Vaginal epithelial cells	Leucocytes	Red blood corpuscles
5 days before heat	Dry	Fairly numerous	Nil	Nil	Dry	Numerous	Nil	Nil	—	—	—	—
4 ,,	—	—	—	—	—	—	—	—	—	—	—	—
3 ,,	—	—	—	—	—	—	—	—	—	—	—	—
2 ,,	Dry	Present	Nil	Nil	Dry	Present	Nil	Nil	Dry	Old present	Nil	Nil
1 ,,	Fluid mucus	Numerous in patches	Few	,,	Moist, fluid mucus	Few	Fairly numerous	,,	Fluid mucus	Old and new numerous	Fairly numerous	Nil
12 hrs. ,,	Moist	Old and new present	,,	,,	Little fluid mucus	,,	,,	,,	,,	Few	Nil	,,
2 ,,	—	—	—	—	Fluid mucus	Several new and old	Several	,,	,,	Several old and new	,,	,,
Beginning of heat	Moist	Few	Very occasional	Nil	,,	Few	Nil	,,	,,	Occasional	,,	,,
2nd hour of heat	Fluid mucus	Very occasional	Nil	,,	Stringy with yellow patches	,,	Numerous	,,	—	—	—	—
4th ,,	Fluid mucus, cheesy lumps	Fluid nil; few	Fluid nil; very numerous	,,	—	—	—	—	Thickish mucus	Few	Few	Nil
8th ,,	—	—	—	—	—	—	—	—	Mucus with cheesy lumps	Several	Very numerous	Nil
12th ,,	—	—	—	—	—	—	—	—	Thick mucus	,,	Numerous	,,
End of heat	Little mucus	Few	None visible	Nil	Mucus with yellow patches	Few	Very numerous	Nil	Little mucus	Several in patches	Few	,,

Table XVIII (contd.).

No. of heifer:	C 1				C 2				C 3			
Period of cycle	Character of discharge	Vaginal epithelial cells	Leucocytes	Red blood corpuscles	Character of discharge	Vaginal epithelial cells	Leucocytes	Red blood corpuscles	Character of discharge	Vaginal epithelial cells	Leucocytes	Red blood corpuscles
2 hrs. after end of heat	—	—	—	—	Thick with yellow patches	Few	Very numerous	Nil	Thick cheesy mucus	Several	Very numerous	Nil
6 "	—	Several	Numerous	Nil	Thick yellow mucus	"	"	"	Thick mucus	Several new and old	"	"
12 "	Thick mucus	Few in patches	Very numerous	"	Little mucus	Several	Few	"	Little thick mucus	Numerous cells in clumps	Numerous	"
24 "	"	Few	"	"	"	Several in patches new and old	"	"	Little mucus	Several old and new	Few	"
36 "	Moist	Numerous in patches	Few	"	—	Several	Fairly numerous	"	Little thick mucus	Numerous in patches	Very numerous	"
2 days after end of heat	Blood tinged mucus	Very few	Nil	Numerous	—	"	Nil	"	Little mucus	"	Very occasional	"
2½ "	"	Several young and old	Fairly numerous	Several	—	"	Few	"	—	Several	Rare	"
3 "	Little mucus	Numerous in patches	Numerous	Nil	—	Several in patches	"	"	—	Few	Nil	"
3½ "	Thick mucus	Several	Several	"	—	Several	Nil	"	—	Several	Rare	"
4 "	"	Few	Few	"	Thick mucus	Few	Numerous	Several	—	Few	Nil	"
5 "	Slight secretion	Fairly numerous	Occasional	"	—	Several old and new	Few	Nil	—	—	—	—
7 "	Dry	Numerous	Absent	"	—	—	—	—	—	Several	Numerous degenerate	Nil
9 "	—	—	—	—	—	—	—	—	—	—	—	—
10 "	Dry	Numerous in patches	Nil	Nil	—	—	—	—	—	Numerous new and old	Occasional	"
12 "	—	—	—	—	Swab red	Numerous in patches	Occasional	Many	—	—	—	—
13 "	Dry	Numerous in patches	Very rare	Nil	—	Numerous	Few	Nil	—	—	—	—
14 "	—	—	—	—	—	—	—	—	—	—	—	—
15 "	—	—	—	—	Dry	Numerous	Few	Nil	Dry	Few	Nil	Nil

these were found to consist of masses of leucocytes with a few vaginal epithelial cells.

After the end of heat, when the flow appeared milky, there was a very large number of leucocytes together with clumps of epithelial cells (Plate XVI, 3).

About two days after the beginning of heat in most heifers red blood corpuscles were added to the flow (Plate XVI, 4).

For another day or two or until about 6 days after the beginning of heat leucocytes still appeared in the secretions (Plate XVI, 5).

After this time and until just before the beginning of the next heat the leucocytes disappeared from the flow and the swabs showed only vaginal epithelial cells (Plate XVI, 6). In some cases, however, it was found that leucocytes persisted for a day or two beyond the 6th day after the beginning of heat. Under some semi-pathological inflamed conditions of the vagina they may occur in all stages of the cycle.

As has been shown above the bulk of the mucous flow is produced by the cervix although a certain amount comes from the upper end of the vagina. In order to determine the origins of the cells in the secretion swabs were taken from different parts of the reproductive tract after the animals had been killed at various stages of the oestrous cycle. Table XIX shows the details of these examinations and it will be seen that although a few leucocytes originate from the uterus and cervix yet the majority come from the vagina just above the urethra and that this area and that of the vagina next the vulva are the main sources of the vaginal epithelial cells. Observations on sections from the vagina (Plate XV) confirm this and the presence of leucocytes in the discharges from about the beginning of heat to about 6 days after it commences synchronises with the period of congestion and its subsidence in the vagina. The uterine epithelial cells are in appearance quite distinct from those of the vagina and they do not appear in the vaginal discharges. Moreover the semipendent position of the uterus in the cow renders it unlikely that its contents would flow easily to the vagina as it does in woman. That the uterus does not add materially to the vaginal discharges is seen by the fact that no red blood corpuscles were found in the cervix as might be expected if the blood flow came from the uterus and although small numbers of red blood corpuscles are found in the uterine mucosa they are mostly absorbed *in situ* giving rise to the amorphous masses of pigment in the process (see p. 88).

To summarise, in the normal animal a few hours before heat smears from the vagina are characterised by the few vaginal epithelial

Table **XIX**. *Cell content of smears from different parts of reproductive tract of heifers killed at various stages of the cycle.*

No. of heifer:		C 5	C 7	C 4	C 6
Stage of cycle:		3 days before heat	On heat	48 hrs. after beginning of heat	72 hrs. after beginning of heat
Uterus		Nuclei of uterine epithelial cells	Nuclei of uterine epithelial cells	Uterine epithelial cells. On cotyledons epithelial cells, few leucocytes and some red blood corpuscles	Nuclei of uterine epithelial cells. Few leucocytes (polymorphonuclear). Several red blood corpuscles
Cervix		Much stringy mucin, nuclei and small patches of cervical epithelial cells	Little stringy mucus. Few epithelial cells of cervix in patches and number of small granules staining with eosin	Thick mucus and patches of cervical epithelial cells numerous	
Vagina	Next os	Stringy mucin. Nuclei of epithelial cells. Occasional leucocyte	Little mucus. Few epithelial cells. Large numbers of leucocytes (polymorphonuclear)	Little mucin. Few epithelial cells. Several polymorphonuclear leucocytes. Large number of red blood corpuscles	Much mucin. Few epithelial cells. Large number of red blood corpuscles and leucocytes
	Above urethra	Clumps of vaginal epithelial cells	Clumps of vaginal epithelial cells and large number of polymorphonuclear leucocytes	Clumps of vaginal epithelial cells. Several polymorphonuclear leucocytes	Little mucin. Few vaginal epithelial cells. Large numbers of leucocytes (polymorph. and eosinophyl), few red blood corpuscles
	Next vulva	Clumps of vaginal epithelial cells	Clumps of vaginal epithelial cells and several polymorphonuclear leucocytes	Clumps of vaginal epithelial cells and several polymorphonuclear leucocytes	Large clumps of vaginal epithelial cells and few leucocytes

cells present owing to their dilution with mucin. For about 6 days after the onset of heat smears from the vagina generally show abundant leucocytes and often red blood corpuscles 2–4 days after the beginning of heat. During the remainder of the cycle only vaginal epithelial cells occur with now and then an occasional leucocyte.

While the utilisation of vaginal smears may be of assistance in helping to diagnose the period of the cycle it is not infallible owing to the frequent occurrence of slight inflammatory conditions in the vagina caused by local irritations.

(g) The Mammary Glands.

The changes undergone by the mammary glands in the course of their development may be conveniently grouped into stages: (1) the development in foetal life, (2) postnatal development up to the age of puberty, (3) growth in the virgin animal after puberty and (4) the changes occurring during the different parts of the oestrous cycle. Later (see p. 168) the changes occurring during pregnancy will also be dealt with.

(1) *The Development in Foetal Life.* Foetuses have been obtained at the end of each "4 week" month of development from the 1st to 8th month of pregnancy. Some were bull calves and others heifer calves but, as will be seen from the description below, there was very little sex difference to be seen in the state of their development. Table XX gives a summary of the changes occurring in the mammary glands during the different months of embryonic life.

The milk line and fold stages of the glands were not obtained as the foetus at 1 month old was required for other purposes.

At 2 months the milk bud of thickened epidermis is formed from the lower layers of the skin epithelium; the top of the bud is level with the general skin surface but the bud itself lies below it. The centre of the bud is slightly depressed and is covered by a layer of rather flattened cells continuous with those of the outer layer of the skin (Plate XVII, 1).

At 3 months the milk bud is elongated and the centre shows a marked depression; the bud is now carried upon the raised dermis which forms the nipple (Plate XVII, 2).

At 4 months the central depression forms a rather long irregular primary sprout which is hollow at its base; this runs in the centre of the nipple and bulges out below forming the cistern. The height of the nipple above the skin surface is increased. Hair buds are present in the general skin surface but the epithelium of the nipple is quite free from these (Plate XVII, 3).

At 5 months the height of the nipple is still further increased and the primary sprout is now well differentiated into a narrow tube at the top of the nipple which forms the stricht-canal and a wide hollow tube below which forms the cistern, at the base of which large secondary lobular buds are being formed. The blood vessels are now very numerous and large throughout the nipple (Plate XVII, 4). In another foetus (from Cow 36) about this period but whose exact age was not known the lobules at the base and side of the hollow cistern were better developed and from them the tertiary sprouts of the minor ducts were being formed.

At 6 months the apex of the nipple is still indented and the top is covered over by the outer or swollen layer of the cutaneous epithelium. The stricht-canal or narrow tube which issues from the apical depression is comparatively long and below it a large cavity, the milk cistern, is seen; a small part of this cavity lies in the nipple itself, *i.e.* above the general level of the skin surface, but the greater part lies below the skin surface or level at which the hair follicles develop. From the cistern walls short secondary sprouts are seen on all sides; these form the main

Table XX. *The stages of development of the mammary gland during embryonic life.*

No. of dam:	P 4	P 3	P 1	P 5	Cow 36	Cow 39	P 7	Cow 35	P 6	P 9
Months pregnant:	2	3	4	5	About 5	About 6	6	About 7	7	8
Sex of foetus:	♀	♀	♀	♂	♀	♀	♂	♂	♂	♂
Nipples:	No nipple. Thickened bud of epithelium below skin surface	Small nipple. Milk bud just beginning to elongate on summit of nipple	Primary growth from depression in nipple; apex forming a hollow duct	Nipple with central duct, no hairs on nipple surface. Blood vessels large and numerous	Nipple hollow. No hair on it. Blood vessels large	As before	Stricht-canal well formed and below it cavity of cistern. Blood vessels large and numerous	As before	As before	Stricht-canal hollow and epithelium with horny layer. Cavity of nipple irregular in cross section, folded and with buds. Numerous large blood vessels
Cistern:	None	None	Duct bulging at bottom and beginning to form cistern	Cistern well formed and beginning to lobulate at lower end	Large cavity and branching into ducts	As before	Well formed sprouts of ducts on all sides	As before. Cistern large	As before	Large, surrounded by network of blood vessels and capillaries
Ducts:	None	None	None	None	Secondary sprouts(main ducts) just forming	Secondary sprouts(main ducts) elongated and tertiary sprouts forming	Secondary sprouts(main ducts) beginning to elongate	Few main ducts sprouting	Main ducts almost reached subcutaneous fatty layer	Main ducts well formed and sprouts of minor ducts (tertiary sprouts) developing just below subcutaneous fat
Alveoli:	None	None	None	None	None	None. Subcutaneous clumps of fat cells	None. Subcutaneous clumps of fat cells	None	None. Subcutaneous fat clumps	None. Clumps of subcutaneous fat

PLATE XVII

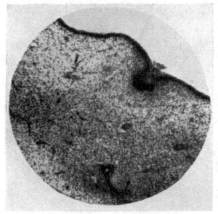

1. Heifer foetus at 2 months.

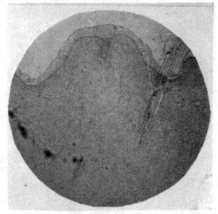

2. Heifer foetus at 3 months.

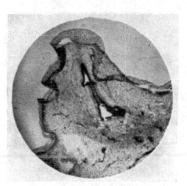

3. Heifer foetus at 4 months.

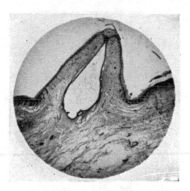

4. Bull foetus at 5 months.

5. Heifer foetus at 6 months.

6. Bull foetus at 8 months.

milk ducts. Small sprouts are to be seen coming from that part of the central cavity which lies in the nipple above the skin level but none appear from the stricht-canal. The blood vessels are large and numerous and are in the main situated in a ringed manner round the growing mammary tissue. In the subcutaneous tissue below this area groups of fat cells can now be observed (Plate XVII, 5).

In another foetus (from Cow 39) at about this period, but whose exact age was not known, a very similar appearance was presented but in addition the tertiary sprouts forming the minor ducts were beginning to appear from the sides of the secondary sprouts or main milk ducts.

At 7 months the appearance was very similar to the last stage but the cistern was more lobulated and the ducts had now almost reached the level of the subcutaneous fatty tissue; round the developing ducts was a mass of blood capillaries. In another foetus (from Cow 35) at about this period, but whose exact age was not known, the ducts were not so well developed although the cistern was large and lobulated. The mouth of the stricht-canal was plugged by a mass of horny epithelial cells.

At 8 months the nipples are large and the stricht-canal is hollow and the epithelium lining it is horny on the outer side. The cistern is well developed and the main ducts (or secondary sprouts) are numerous, they are beginning to show the tertiary sprouts of the minor ducts and have almost reached the subcutaneous fatty layer. The blood vessels are large and numerous and form a network round the nipple and cistern; capillaries are very numerous round the cistern and milk ducts in much the same way as they surround the base of the hair follicles (Plate XVII, 6).

The above account of the development of the mammary glands in the cow is on the whole similar to that given by Profé (225) in which the age of the foetus is not stated but only its length. We do not however agree in some of the details; for example, we have observed no case of a swollen mammary bud extending to the base of the nipple (his Fig. 16) but have found that the milk bud begins to elongate as the nipple is formed (Plate XVII, 2) and the first down growth from it appears as a narrow tube which soon becomes hollow at its lower end (Plate XVII, 3). The mammary pouch or bud which first appears (Plate XVII, 1) forms the plug of epithelium in the mouth of the stricht-canal, that is, the whole original pouch does not grow down, enlarge and then become hollow but only its lower layer dips down to form a tube—the primary sprout; the nipple itself develops as a growth of the dermis which carries up the milk bud on its summit. We agree with him that the primary milk sprout from the bud develops at the same time as the hair buds appear in the skin and that the duct system of the cow's nipple is not (as Gegenbaur and Klaatsch describe) the persistent mammary bud but is formed (as Rein [226], Curtis and Tourneux describe) from the primary sprout (see Bonnet [227] and Bresslau [228]). Lustig [229], in woman, figures many and not one primary sprout from the milk bud.

Zschokke [230] has recently described the development of the duct system of the mammary gland in cattle for foetuses of known lengths and in a few cases of known ages. He also agrees that the hair follicle formation starts at the same time as the primary sprout of the mammary gland. Since the teat itself is devoid of hair sprouts it is reasonable to suppose that in the growth of the nipple the mammary bud is partially expended in the formation of its epithelium and that the epithelium of the nipple is not formed by the unspecialised skin epithelium which produces hair follicles. We believe that not only does the mammary bud form an area at the apex of the nipple as Zschokke figures but that it also supplies the hairless part of the nipple epithelium.

We agree with him that the primary sprout forms the stricht-canal of the teat (Plate XVII, 3 and 5) and that the lower lobulated portion of this forms the cistern. In Plate XVII, 5 the comparative length of the stricht-canal appears large but, since the growing area of the nipple is its base and not its apex, this part becomes relatively small in the later stages of development (new-born and heifers). There is no evidence that the central hollow of the adult nipple is formed from this narrow portion but rather that in development it is produced from the first formed or apical part of the cistern. This would explain why rudimentary alveoli are frequently found in the walls of the adult nipple cavity as Riederer [143], Ernst, Mohler and Eichhorn [231] have described and we have also observed. The large milk ducts are the secondary sprouts formed from the lower (hollow) part of the primary sprout. The complicated network of blood vessels in the nipple is seen to be developed at an early stage.

(2) *Postnatal Development up to the age of Puberty.* An anatomical description of the glands of calves from 2 to 6 months old has already been given above. Histological examination of these showed that the gland consisted of a milk cistern with a very small cavity from which a branching system of ducts had developed. No true alveoli had been formed but the ends of the branching ducts were rather swollen (Plates XVIII, 1 and XIX, 1). Included under this heading may be taken the case of the freemartin (No. 38) which (see p. 67) was three years old

PLATE XVIII

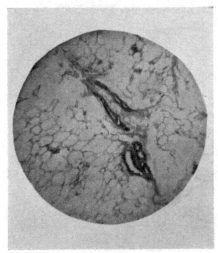

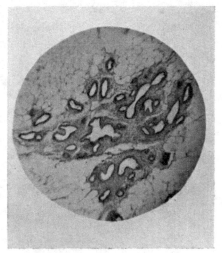

1. Mammary gland of calf
6 months old.

2. Mammary gland of freemartin
3 years old.

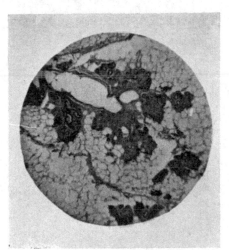

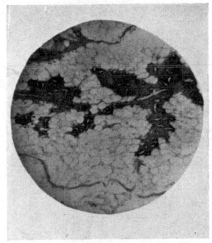

3. Mammary gland of heifer about
14 hours before oestrus.

4. Mammary gland of heifer 8 days
after oestrus.

and had a very large udder but a very small amount of mammary tissue. Histologically the gland was in the same stage of development as that of calves of 6 months old, namely, with branching ducts but with no alveoli (Plates XVIII, 2 and XIX, 2).

(3) *Growth in the Virgin Animal after Puberty.* Although to external appearances the udder of the freemartin quoted above was as large or larger than any of the virgin heifers examined (see p. 67) yet on staining the mammary tissue it was seen to be much less developed, the ducts of the heifers after puberty being about five times as long and the size of the milk cistern in them being much larger (see Plate V, 1 and 2). Histological examination showed that while in the freemartin no alveoli had formed yet in virgin heifers after puberty a moderate quantity of alveolar duct tissue was present (Plates XVIII, 3 and XIX, 3). Thus between the calf which has not ovulated and the heifer which has ovulated there is a great difference in the appearance of the gland (compare Plate XVIII, 1 and 3); the former consists of ducts only while the latter contains marked lobules of alveolar ducts. Ernst, Mohler and Eichhorn (231) describe the gland in the new-born calf as consisting mainly of cell tubes and buds embedded in connective tissue rich in fat; the ends of the tubes are dilated but no true alveoli are present; at puberty the alveolar ducts appear and are surrounded by strong connective tissue. The ducts of the gland as they grow develop in and along the thick bands of connective tissue in the udder, these are separated by masses of fat cells.

(4) *The Changes occurring during Different Stages of the Oestrous Cycle.* Since at this stage of its development the area that the gland covers is a large one differences will be found in the histological appearance of small pieces taken from different parts of the gland, whether from near the milk cistern or from the branching upper ends of the ducts.

If observations are confined to the lobules of alveolar ducts and more especially at their growing ends, changes can be seen in the alveolar ducts and their epithelium at different stages of the cycle (see Table XVI). Just before heat begins the lumen of the alveolar duct is large and filled with secretion and the epithelium is cubical (Plates XVIII, 3 and XIX, 3), whereas at 8 days after heat the lumen is small and the cells of the gland are almost columnar in shape (Plates XVIII, 4 and XIX, 4); and between these all gradations exist.

It would seem as though the gland underwent a cycle of changes during the oestrous cycle corresponding to but lagging behind that of the corpus luteum, the mammary epithelium being enlarged during its development and shrivelling during its involution. As has been pointed out above (see p. 68) the amount of the secretion drawn off from the nipples is not likely to vary so much as the character of the epithelium would suggest since the cistern acts as a reservoir. The chemical composition of the secretion would suggest that the fluid is not altogether a result of secretion by the epithelial cells but consists in part of a filtration of fluids from the blood tending to fill the cistern. The presence of casein fat and lactose in the secretion at this stage (see Woodman and Hammond(169)) points to the activity of the alveolar ducts which histological examination shows to vary at different periods of the cycle.

On the whole the histological changes in the mammary gland during the oestrous cycle are slight compared with the changes before and after puberty and also before and after pregnancy. The individual variation in development is also so great that the transient effect of the corpus luteum on the mammary gland during the cycle is often obscured.

A comparison of the above described changes in the mammary gland of the cow with those that occur in the rabbit during pseudo-pregnancy (see Hammond and Marshall(60)) shows that the large development of ducts and alveoli which occurs in the rabbit during pseudo-pregnancy corresponds to the growth of the gland in the heifer at puberty when for the first time the corpus luteum is formed in the ovary. Since in the heifer another corpus luteum appears before the last has completely atrophied no great involution of the alveolar ducts and main ducts occurs as it does in the rabbit where a long period may elapse before another corpus luteum is formed.

Myers(232) found similarly in the rat that great activity and growth occurs at puberty when the differences between the mammary glands of males and females become marked.

Loeb and Hesselberg(233) have studied the cyclic changes in the mammary gland of the guinea-pig and concluded that the corpus luteum was responsible for the secondary growth changes; they state however that the absence and not the presence of the corpus luteum was associated in point of time with the growth of the mammary gland, the maximum development of the mammary gland occurring at the time of heat and the minimum some 6 days after oestrus. The cycle in the guinea-pig however is short and in the rabbit (see Hammond and Marshall(60)) it is at least 5 days before the effect of the corpus luteum is seen on the growth of the gland so that, allowing for this delay period, it will be seen that the cycles in the ovary and mammary gland correspond. The effect of the corpus luteum on the mammary gland is

PLATE XIX

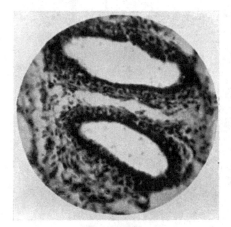

1. Calf: 6 months old.

2. Freemartin: 3 years old

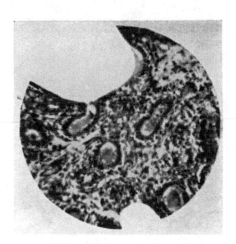

3. Heifer: 14 hours before oestrus.

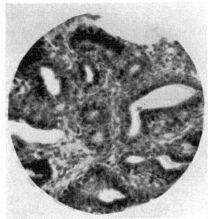

4. Heifer: 8 days after oestrus.

THE MAMMARY GLAND.

well shown in the monoestrous dog (Marshall and Halnan [73]) where the following ovulation and formation of a new corpus luteum takes place at a long interval.

Lenfers [234] has described the histological appearance of the udder of the cow before and during pregnancy and also during lactation and when dry; a detailed account is also given by Martin [235]. Our observations show that the histological appearance of the udder of the cow varies considerably not only with the period of lactation but also with individuals and whether or not the alveoli are empty after milking or full before it. These variations are so great that it would be useless to look for cyclic variations in the udders of cows which have produced a calf.

IV

PREGNANCY

Introduction.

Oestrus and the rhythm of the reproductive organs are normally suspended during pregnancy and although we have been informed of cases of oestrus occurring occasionally during almost all stages of pregnancy (see Appendix I (9)) no such cases have come under our direct observation. Schmid (16) found that heat might occur in the cow at all stages of pregnancy. Whether in these cases ovulation actually occurs must be doubted, although Williams (236) believes that it does occur and that pregnancy is not interrupted but the newly formed corpus luteum takes over the function of the corpus luteum of pregnancy. Since removal of the corpus luteum during pregnancy usually causes abortion it is difficult to see how this can happen, unless in the cow the formation of a new corpus luteum takes place so rapidly that the absence of its secretion is not interrupted; but this is doubtful. Strodthoff (59) from the rectal examination of cows' ovaries concluded that an incomplete cycle occurs during pregnancy but that the follicles usually became atrophic instead of ripening. He states that frequently a follicle does ripen during pregnancy and that the cow comes on heat; 9 cases are given of cows pregnant from 24 to 114 days which came on heat. Hess (41) gives the case of a cow which came on heat when pregnant and in which a small cyst was found in the ovary near the corpus luteum. In some animals, such as the rabbit, coitus normally occurs during pregnancy (Hammond and Marshall (24)) but ovulation never takes place and it may be that these cases in the cow are similar.

There is a wide belief in practice that heifers which are in the early stages of pregnancy fatten more easily than those which are not pregnant. This is probably due to the fact that in these animals oestrus with consequent excitement and consumption of energy does not occur.

The changes which occur in the reproductive organs during pregnancy have been studied by killing heifers previously virgins, all about the same age, at the end of each 4-week period of pregnancy from the 1st to 8th month (Heifers P 1–P 9).

In addition to these, material was collected from slaughter-houses and the stage of pregnancy estimated from the size of the foetus in comparison with these known stages.

The weighings, observations and treatment of the material were similar to those described above, made on animals killed at different stages of the oestrous cycle and all figures given for size are comparable.

The changes, both anatomical and histological, will be described in various sections below. Owing to limitations of time, space and money no experimental work has been done on this part of the subject; it is one which has received but little attention although its economic bearing on the nutrition of the young and abortion is considerable.

The Diagnosis of Pregnancy. Clinically the diagnosis of pregnancy in the cow is important more especially in the early stages since in order to get cows calving about the same time each year there is not much time to be lost and the missing of one or two heat periods may disturb the supply of milk at a time when it is wanted.

Diagnosis can be made from the result of a number of different observations, each considered separately being liable to error for some reason or another as will be shown below, but together they form a reliable method.

The first symptom and the one which is generally used is the absence of heat at the expected time; this however is not infallible as heat may be prevented by the persistence of the corpus luteum (see Sterility, p. 184) or by conditions being unfavourable (such as tuberculosis) for the development of the follicle (see Cow A 4, p. 183) or may be so short that its occurrence is overlooked (p. 18).

The next symptom that is most common in general use is that of palpation of the foetus through the body wall by pushing in the right flank with the hand. This, however, can only be used in the later stages of pregnancy after about the 5th month and is not infallible as occasionally the calf may lie on the left side over part of the rumen when it is not felt and also a uterus filled with fluid gives much the same touch as an early stage of pregnancy. Sand[237] found that in 6 out of 11 cows the pregnant uterus was lying next the abdominal wall outside the omentum; in this position it can pass under the rumen and lie in the left side of the cow and in one of the 6 cows it was found in this position although it had been lying on the right side the day before. When the uterus lies within the omentum this cannot happen. In the heifer killed at the 8th month of pregnancy the movements of the foetus could be seen when the right flank was carefully watched. Experienced herdsmen state (see Appendix I (11)) that the foetus can be felt at 6 months at which time movements can be felt when the cow is being milked and some say that it can be determined as early as 4 months when no food is present in the stomachs.

Diagnosis by examination of the ovaries per rectum is possible only in the early stages (1–2 months), after this time the weight of the uterus pulls the ovaries down deep into the body cavity and they are not easy to handle. This examination requires care as injury to the corpus luteum is liable to be followed by abortion. If a large corpus luteum is present in the ovaries at the time when the first or second heat period after service is due there is a reasonable supposition that the animal is pregnant, although a persistent corpus luteum has to be considered.

Albrechtsen[120] states that it is possible to irrigate the pregnant uterus of a cow without causing abortion and has seen some 100 cases treated in this way up to $3\frac{1}{2}$ months of pregnancy. Oppermann[238], however, found that irrigation of the uterus in pregnant cows causes abortion in 24 hours. Examination of the uterus per rectum forms, especially in heifers, a very easy and safe method of diagnosing pregnancy both in its early and late stages. In heifers the uterus normally lies well within the pelvic cavity; in cows, and more especially old ones in which the ligaments have become relaxed, the uterus hangs down further into the body cavity over the edge of the symphysis pubis. The normal position of the uterus in relation to the pelvis is well shown in figures given by Oppermann[239]. As diagnosis of pregnancy per rectum depends in the early stages on the position, size and consistency of the uterus it is important to bear in mind the age of the cow when making an examination, as differences exist in these points between heifers which have not produced a calf and old cows. It is believed too, although no systematic examination has been made, that the uterus is more mobile in the milk breeds (Friesian) than in the beef breeds (Shorthorn). Table XXI shows the results of the examination of the uterus per rectum from the 3rd to 8th month of pregnancy in a series of heifers in calf for the first time. The first indication of pregnancy is the unequal size of the two horns of the uterus accompanied by a soft touch of the larger side as opposed to the firmer feel of the non-pregnant horn or uterus. Albrechtsen[120] states that in the first 5 or 6 weeks diagnosis cannot be made absolutely. After this time, however, the pregnant horn becomes larger than the other one and in the latter part of the 2nd month pregnancy can be safely recognised by the palpation of a fluctuating oblong body in the larger horn. Zieger[110] says that a difference in size of the two horns can be observed at the beginning of the 2nd month. By the 3rd month the uterus has so enlarged as to hang down into the

Table XXI. *Diagnosis of pregnancy by examination per rectum of heifers in calf for the first time.*

No. of heifer:	P 3	P 1	P 5	P 7	P 6	P 9
Stage of pregnancy (mths.):	3	4	5	7	7	8
Uterine horn—position	Dropped into body cavity	At edge of symphysis pubis	Sunk in body cavity. Division between horns not felt	Sunk in body cavity	Dropped in abdominal cavity	Dropped in abdominal cavity
Uterine horn—consistency	—	Soft	Soft	Soft	Soft	Soft
Uterine horns—size	—	Swollen	Large	Large	Very large	Very large
Cotyledons	—	Not felt	Not felt	Could be felt	Could be felt	Could be felt
Foetus	—	,,	,,	Not felt	Outlines not felt	Limbs and movements felt through uterus
Uterine artery	—	—	—	—	Pulsation distinct	—
Ovaries	One ovary only felt	Not easy to feel	Not felt	Not felt	Not felt	Not felt
Cervix	—	Distinct from uterus	Distinct from uterus	—	Not very distinct	Not very distinct

body cavity, falling over the edge of the symphysis pubis and carrying the ovaries with it so that these are difficult to find more especially on the side which is pregnant. By the 5th month the uterus has so increased in size that the line of division between the two horns of the uterus is quite out of reach as also are the ovaries. Sand (237) states that although at the 4th month the foetus can be palpated per rectum yet by the 5th month the uterus has sunk down into the body cavity so that it is impossible to feel it. By the 6th month the cotyledons could be distinctly felt through the wall of the uterus; Albrechtsen (120) states that the cotyledons can be first felt at the 3rd month. During the 7th month the limits of the cervix are not distinct from the body of the uterus but it appears to be pulled out by the weight of the pregnant uterus. Williams (236, p. 141) states that in the latter part of pregnancy the cervix may be pushed back into the vagina to such a degree that it may even appear between the lips of the vulva when the animal is lying down. Zieger (110) says that from the 7th month of pregnancy, on account of the pull of the uterus, the cervix is drawn forward in front of the os pubis but towards the end of gestation the cervix is forced back again into the pelvic cavity. Reinhardt (240), however, figures the cervix in front of the os pubis just before birth. By the 8th month the limbs of the foetus and its movements can easily be felt through the wall of the uterus.

The accumulation of fluid in the uterus or the presence of a mummified foetus are circumstances liable to lead to error in diagnosing pregnancy by this method.

Williams (236, p. 157) has described changes in the pulsation of the uterine artery which can be felt against the angle of the pelvis (see figure by Oppermann (239)). The only observation made in this series—at the 7th month of pregnancy—showed the pulsations to be very marked and distinct at this time.

A very useful indication of the presence of pregnancy in both cows and heifers is the nature of the mucous secretions from the os. Although, as has been shown above (p. 54) the character changes at different periods of the oestrous cycle, yet the quantity is comparatively small compared with that during pregnancy, the amount in the cervix increasing as pregnancy advances. The mucous secretion during pregnancy is always very thick and sticky and never fluid. It can be sampled by taking some from the cervix between the finger and thumb or can be obtained by hooking the os and pulling it back into the vagina and inserting a wire vaginoscope. Williams (241) found the cervical seal to be well formed at the 30th day of pregnancy.

Oppermann (239) states that during pregnancy the vaginal walls are dry and more sticky than during the cycle and it is more difficult to insert the hand.

The vaginal secretions do not form a certain guide as to the presence of pregnancy but absence of blood from the secretions in these animals which normally bleed after heat periods and the absence of numerous leucocytes from the secretion at a time after the next oestrus would have been due are indications that pregnancy may be expected. Local inflammations however and the occurrence of granular vaginitis (Wilson (242)) are liable to upset diagnosis based on these grounds.

The nature of the fluid obtained from the nipples of previously virgin heifers in calf for the first time also serves as an indication of pregnancy, as will be seen from Table XXII. The fluid obtained up to the 4th month of pregnancy and also in the non-pregnant state is serous and very liquid; during the 5th month of pregnancy the consistency becomes thick and honey-like and of yellow colour, drying on the hands in white flakes. At the 7th month a secretion having the appearance of rich milk or colostrum was obtained but this heifer had been milked out when in the honey-like secretion stage. It is uncertain whether this appearance was produced normally or as a result of the previous milking, probably the latter, since the heifer killed at the end of the 8th month

Table XXII. *Nature of secretions from the nipples at different periods of gestation in heifers pregnant for the first time.*

No. of heifer	Before pregnancy	Month of pregnancy							
		1	2	3	4	5	6	7	8
P 1	—	—	—	—	Serous fluid secretion	—	—	—	—
P 2	—	Serous fluid secretion	—	—	—	—	—	—	—
P 3	—	—	—	Serous fluid secretion	—	—	—	—	—
P 5	Serous fluid secretion. On standing white ppt. and clear yellow fluid	—	Serous fluid secretion	Serous fluid secretion	Serous fluid secretion	Viscid fluid from fore-quarter	—	—	—
P 6	Serous fluid secretion	Serous fluid secretion. Under 20 c.c.	Serous fluid secretion	—	Serous fluid secretion 45 c.c.—yellowish fluid and cloudy ppt. on standing	—	Very viscid honey-like secretion. Light yellow colour. Over 100 c.c. Dried on hands in white flakes. 1 quarter with fluid secretion	Rich milk-like fluid—layer of cream on standing 280 c.c. from two quarters	—
P 7	Serous fluid. On standing clear yellow fluid and white ppt.	Serous fluid secretion	Serous fluid secretion	—	Serous fluid secretion	—	As above. One quarter with fluid secretion	—	—
P 9	—	—	—	—	—	—	—	—	Very viscid honey-like secretion from all nipples

and which had not previously been milked still showed the honey-like secretion such as occurred at the 5th or 6th month. Herdsmen have informed me (see Appendix I (11)) that even in cows in milk the nipples have a sticky feel when pregnant 3–4 months but no direct observations have been made on this point.

Schmaltz[1] has described very fully the means of ascertaining whether a cow is pregnant or not and Williams[236, p. 150] also gives a detailed account of the methods of diagnosis of pregnancy in the cow; he states that in heifers distinct enlargement of the uterus occurs within 20–30 days after conception but that in cows it cannot be observed until 30–70 days after service. He believes that the best time to diagnose pregnancy is 18–20 days after service when the cow if in calf will have a large corpus luteum of pregnancy but if not pregnant then no corpus luteum can be felt in the ovaries.

Duration of Pregnancy. No direct observations have been made on this point but enquiries were made from herdsmen (see Appendix I (10)) who state that the average time is 40 weeks but that with heifer calves it is slightly shorter and with bull calves slightly longer. Spencer[243] found in Shorthorn cattle that the largest number of cow calves were born on the 284th day and bull calves on the 285th day. Most authorities agree that the duration varies considerably; Franck-Albrecht[244] give a very complete account of the literature on this point. Wellmann[245] found that Hungarian cows averaged 285 days whereas Simmental cows averaged 291 days and Wilhelm (see [244]) states that the average period of gestation in the Hungarian cow is some ten days longer than in the Dutch cow, that is, the period of gestation is shortened in the improved breeds.

Ewart[246] states that in the horse the coarse types have a pregnancy of 356–359 days whereas in the finer breeds it is only 334–338 days. Since the duration of the oestrous cycle is reduced by better nutrient conditions (p. 11) it may be that nutrition acts in a similar way in both cycle and pregnancy by affecting the life of the corpus luteum. Marshall[247] states that in dogs the smaller breeds tend to have a somewhat shorter period of gestation than the larger but Smith[248] believes that this is not so. Lefour (see [243]) states that the period of gestation is longer in the larger German breeds of cattle than in the smaller ones. Wellmann[245] found that in mares there is a difference in the length of the gestation period according to the season of the year; pregnancy averaged 322 days when the births occurred in July and there was a constant increase up to May when the average duration was 346 days;

he also found a similar variation in cows but to a smaller extent. Since it has been shown (p. 11) that the length of the oestrous cycle varies with the time of year, probably through the life of the corpus luteum, it is not unreasonable to suppose that similar effects are caused in pregnancy. Williams (236) states that the normal variation of the duration of pregnancy in the cow is 20–30 days on the average duration of 280 days. He shows that the variation cannot be due to the time of oestrus at which service takes place as it only lasts a short time; moreover the egg is shed after oestrus itself ends. He believes that the variation is due to the fact that the foetus can be expelled in a state of relative immaturity; while this is no doubt true it does not account for the variation beyond the average time of 280 days. Periods of 321 days and more have been recorded by various writers (see Franck-Albrecht (244)). Berry-Hart (249) believes that in cows variation in the duration of pregnancy is caused by the spermatozoa injected at one heat period living and fertilising the egg shed at the next period, but the variation curves which he gives do not support this view for there are no peaks at 3-week intervals. He quotes data collected by Krahmer which show that while the gestation periods of one individual may vary there are also large individual differences; in 8 cows the longest individual average was 302 days and the shortest 277 days. In woman where coitus is allowed at any period of the cycle the duration of pregnancy might be affected by the period between insemination and fertilisation (see Triepel (70)) but this is only a few hours in the cow.

It has been shown above (p. 9) that the variation in the length of the cycle due to the duration of the corpus lutuem may be normally about 5 days in a cycle of 20 days; since the duration of the corpus lutuem in pregnancy is 14 times as long it will be seen that the variations in both pregnancy and the cycle are of the same degree. No opportunity has occurred for testing whether cows which have abnormally long or short oestrous cycles have also abnormally long or short gestation periods; Vignes (188) gives figures for women which indicate that this is so, cycles of 30–40 days averaging a pregnancy of 288 days whereas cycles of 21–26 days averaged a pregnancy of 269 days. There is much room for experimental investigation on this point and in determining whether the development of the foetus or the corpus luteum regulates the length of pregnancy and is the cause of birth or whether the pituitary plays any part in the process. A knowledge of these facts would be of assistance in combating abortion.

Parturition. No detailed study has been made of parturition in the cow but it has been observed incidentally and the symptoms of approaching parturition have been ascertained from experienced herdsmen (see Appendix I, (12)). They say that the vulva swells usually about a week but sometimes up to 6 weeks before calving; the ligaments relax and the skin drops on each side of the tail head between the pin bones (relaxation of sacro-sciatic ligament) from 24 hours to a few hours before calving; a flow of slime takes place from the vulva anything from 4 days to 3 hours before parturition and is occasionally seen some two months before; the udder swells markedly about a week before calving and just before parturition takes place, the cow appears uneasy and restless and stands apart from the others, while urine is passed rather frequently.

Zieger[110] found that the pelvic ligaments began to sink as much as 4–6 weeks ante-partum and became normal again about 8–12 days postpartum; he found that pregnancy could be detected by auscultation of the foetal heart-beat which varied from 112 to 150 as compared with the mother's 75–85 rate. The relaxation of the sacro-sciatic ligament before parturition has been described by Fuhrimann[250] who points out that it also occurs under other conditions such as cystic ovaries and consists of a serous infiltration. It will thus be seen that serous infiltration and swelling occur in the pelvic ligament, vulva and mammary gland just before parturition; since the same symptoms occur in cases of nymphomania (cystic ovaries) it must be concluded that all are due to stimuli of ovarian origin. The literature on the relaxation of the pelvic ligaments has been summarised by Schmaltz[1].

Limmer[251] and Stapel[252] both found a slight rise in body temperature of the cow during gestation and a drop just before calving. Weber[253] states that the average rise is 1·6° and that the drop begins from 52 hours to 15 hours before calving.

It will be seen that the symptoms of parturition are very similar but more intense than those which occur at the end of pseudo-pregnancy or during a normal heat period. It is possible that these sets of symptoms are caused by the reduction in quantity of the ovarian hormone or atrophy of the corpus luteum and that the difference in degree between pro-oestrus, the end of pseudo-pregnancy and parturition is due to the duration of the trophic action of the ovarian hormone. Abortion and symptoms similar to parturition are produced by removing the corpus luteum during pregnancy. In the rabbit the end of pseudo-pregnancy which coincides with the atrophy of the corpus luteum is often followed as in true pregnancy by the pulling out of fur and making of a nest (see Hammond and Marshall[24]).

An objection to this view, however, is that the cow does not usually take the bull and the follicle does not ripen until about 30 days after calving, but it may be that desire for coitus is distinct from the other objective signs, such as swelling and bleeding, and that a degree of involution of the corpus luteum which is not sufficient to supply the trophic needs of a large uterus and its contents is still more than sufficient to prevent the ripening of the follicle (compare menstruation and ovulation in women, p. 27).

(a) The Ovaries.

The Corpus Luteum. The chief interest in the ovaries of pregnant heifers lies in the corpus luteum. As will be seen from Tables XXIII and XXIV the corpus luteum in a pregnant animal persists at maximum size throughout gestation (see Plate XX, 1 and 2) instead of disappearing as in the cycle after a life of three weeks. Moreover it will be seen that the diameter of the corpus luteum of pregnancy is slightly, but not much, more than its maximum development in non-pregnant heifers of the same age, being 2·0 cm. in the latter (variation 1·9–2·1) and 2·1 cm. in the former (variation 2·0–2·3). Its small but not marked increase in size during pregnancy suggests that its life is prolonged by maintenance of favourable conditions of life rather than direct stimulus to growth by the condition of pregnancy. Although the actual size of the corpus luteum does not change, during pregnancy, its form alters slightly; at first it protrudes from the ovary but later tends to sink into the ovarian stroma and becomes more rounded instead of oblong. The colour— a yellowish orange—remains much the same throughout pregnancy without marked changes such as occur during the cycle; the colour is much the same as that of the periodic corpus luteum at mid-cycle, as has been pointed out by Küpfer[89].

In histological appearance too the main difference between the corpus luteum of pregnancy and the periodic corpus luteum is that the former maintains its organisation throughout pregnancy up to at any rate the 8th month (and the investigation has not been taken further) and shows no change either in size (see Tables XXIII and XXIV) or structure (see Plates VIII, 6 and X, 4), its appearance being almost identical with that of a corpus luteum during the mid-period of a cycle (Plates VIII, 2 and IX, 4). On the other hand the periodic corpus luteum undergoes atrophic changes three weeks after its formation.

Several investigators have examined corpora lutea to find out whether any differences existed in the histological structure of the corpus luteum

Table XXIII. *Size and weight of parts of the ovaries of heifers pregnant for the first time.*

		P 2	P 4	P 3	P 1	P 5	P 7	P 6	P 9
No. of heifer		P 2	P 4	P 3	P 1	P 5	P 7	P 6	P 9
Month of pregnancy		1	2	3	4	5	6	7	8
Approximate age (yrs. and mths.)		2–6	2–0	2–6	2–6	3–0	3–0	2–6	3–0
Weight of ovary (gm.)	Ovary with corpus luteum	10·9	10·1	8·65	7·3	7·9	10·0	11·0	7·0
	Ovary without corpus luteum	5·3	3·7	3·55	4·3	3·35	3·9	5·9	2·8
Diameter of corpus luteum (cm.)	C.L. of pregnancy	2·3	2·0	2·3	2·0	2·05	2·13	2·0	2·13
	C.L. of previous ovulation	0·6	—	—	—	—	—	—	—
Diameter of Graafian follicle (cm.)	Largest	1·1	1·1	1·2	1·3	1·1	0·8	0·4	1·0
	Next largest	0·5	0·9	0·4	0·8	0·8	0·8	0·4	0·7

Table XXIV. *Size and weights of parts of the reproductive organs of pregnant cows and heifers.*

No. of animal		42	41	18	39	43	36	35
Approximate stage of pregnancy (mths.)		1	3½	3½	4½	4½	5	6
Approximate age		Heifer	Heifer preg. 1st time	2nd calf	1st calf	1st calf	1st calf	Cow
Ovary (weight gm.)	Ovary with corpus luteum	12·0	7·7	10·7	7·3	8·5	7·0	9·9
	Ovary without corpus luteum	4·0	3·0	5·7	3·8	4·8	4·2	3·1
Corpus luteum (diameter cm.)		2·4	2·2	2·3	2·3	2·1	2·3	2·1
Graafian follicle (diameter cm.)	Largest	1·0	1·1	1·3	1·1	1·1	1·1	0·7
	Next largest	0·9	0·7	0·3	0·4	0·5	0·4	0·6
Uterus and contents (weight gm.)		—	—	4,200	7,000	—	48,400	121,000
Foetus (weight gm.)		3	340	560	1,010	1,100	2,360	11,925
Mammary gland (weight gm.)		—	—	2,600	—	4,580	1,440	—

of pregnancy and that of the periodic corpus luteum. Ravano (254) could find no essential difference in women and Marcotty (255) also found in women that the corpus luteum of pregnancy remains in the well developed premenstrual state of the periodic corpus luteum; he describes a few minor differences such as the greater distinctness of the central plug and the more frequent occurrence of calcareous concretions in the corpus luteum of pregnancy but we are unable in heifers to distinguish with any certainty between the two types. Hammond (256) found in the rabbit that the corpora lutea of pregnancy slightly exceeded in size those of pseudo-pregnancy and Corner (107) in sows found the corpora lutea of pregnancy to be slightly larger than that of the periodic corpus luteum. Marcotty (255) states that in women the corpus luteum of pregnancy only equals in size the maximum stage of the periodic corpus luteum.

In heifers up to the 8th month of pregnancy no sign of degeneration exists in the corpus luteum and since after parturition a cow does not

PLATE XX

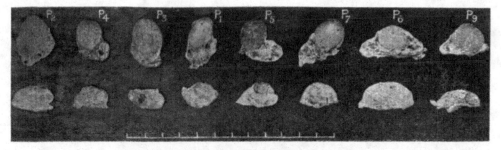

1. Ovaries of heifers killed during pregnancy.

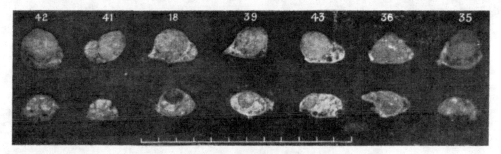

2. Ovaries of pregnant cows from slaughterhouses.

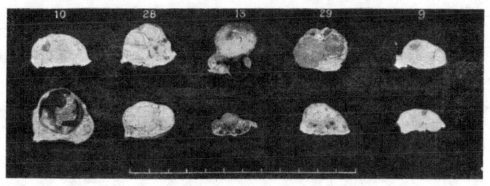

3. Ovaries of sterile cows.

normally come on heat for about 30 days it may be concluded that the atrophy of the corpus luteum is delayed until that time. Delestre (100) states that in the cow degeneration of the corpus luteum begins after the 5th month of pregnancy; he based his conclusions however only on histological examination and made no measurements of the size of the corpus luteum. Schmid (16) describes degeneration (histologically) of the corpus luteum as early as the 2nd month of pregnancy but the measurements of size that he gives show no decrease during pregnancy (see Table XXV, *e*). From his plates it appears to us that he has taken connective tissue cells for degenerate luteal cells. Zietzschmann (26) states that the corpus luteum of the cow during pregnancy maintains its maximum development for a long time and until just before parturition. Küpfer (89) who made detailed measurements of the corpora lutea of cows during the cycle and during pregnancy concluded that the corpus luteum of pregnancy was not larger than the periodic corpus luteum and gives two cases of non-pregnant animals at the 16th day of the cycle in which the corpora lutea were larger than those of any pregnant animals. He states that the corpus luteum is largest during early pregnancy and becomes slightly reduced in size towards the end, but his figures show considerable variation and were obtained from animals of different breeds which we consider would account for the variations in size observed. He states that after parturition the corpus luteum of pregnancy degenerates in the same way as does the periodic corpus luteum and the cow ovulates again 16–21 days after calving. Corner (107) also found in the sow that the degeneration of the corpus luteum after parturition was essentially the same as that which takes place in the periodic corpus luteum at the end of the cycle. Williams (84, p. 262) states that the corpus luteum of the cow remains at full size up to and through parturition and does not normally atrophy until 30–60 days after the termination of pregnancy when she is again on heat. Direct weighings of corpora lutea at different known stages of pregnancy have been made in the cow by Bergmann (257) and are shown in Table XXV (*c*); from these it will be seen that full size is maintained until the end of pregnancy. Hess (41) states that heat does not normally occur after parturition until such a time (6–8 weeks) as the corpus luteum atrophies.

The weight of the corpus luteum may be estimated indirectly by comparing the weight of the ovary which contains the corpus luteum with the other one without this body. The ratio between the two is shown in Table XXV (*a*) for pregnant heifers and it will be seen by comparing these figures with those given in Table IX for non-pregnant

Table XXV. *Weights and relative size of two ovaries in pregnant cows and heifers.*

(a) Heifers of known stages of pregnancy.

Stage of pregnancy (mths.)	1	2	3	4	5	6	7	8
No. of animal	P 2	P 4	P 3	P 1	P 5	P 7	P 6	P 9
Sum weight of two ovaries (gm.)	16·2	13·8	12·2	11·6	11·2	13·9	16·9	9·8
Ratio $\dfrac{\text{Ovary with corpus luteum}}{\text{Ovary without corpus luteum}}=100$	206	273	243	170	236	256	186	250

(b) Heifers and cows of estimated stages of pregnancy (from weight of foetus).

Stage of pregnancy (mths.)	1	3½	3½	4½	4½	5	6
No. of animal	42	41	18	39	43	36	35
Sum weight of two ovaries (gm.)	16·0	10·7	16·4	11·1	13·3	11·2	13·0
Ratio $\dfrac{\text{Ovary with corpus luteum}}{\text{Ovary without corpus luteum}}=100$	300	257	188	189	173	167	319

(c) Heifers and cows of known stages of pregnancy (from Bergmann [257]).

Month of pregnancy	2	3	4	5	6	7	8	9	10
No. from which calculated	3	6	13	7	4	3	5	5	2
Sum weight of two ovaries (gm.)	12·0	18·1	17·6	19·1	20·0	19·6	17·3	14·6	11·8
Ratio $\dfrac{\text{Ovary with corpus luteum}}{\text{Ovary without corpus luteum}}=100$	224	148	171	148	126	180	198	204	237
Weight of corpus luteum (gm.)	4·5	4·7	5·5	4·9	4·8	5·8	4·9	5·1	5·2

(d) Heifers and cows separately of estimated stages of pregnancy (from lengths of foetus) (from Küpfer [89]).

		1	2	3	4	5	6	7
Month of pregnancy		1	2	3	4	5	6	7
No. from which calculated	Heifers	4	4	5	2	1	2	1
	Cows	2	1	1	1	1	5	3
Sum weight of two ovaries (gm.)	Heifers	15·5	15·0	15·8	14·7	12·7	12·2	12·3
	Cows	20·9	23·8	16·5	14·1	23·1	24·7	19·9
Ratio $\dfrac{\text{Ovary with corpus luteum}}{\text{Ovary without corpus luteum}}=100$	Heifers	206	211	243	198	217	184	242
	Cows	175	183	150	182	165	181	201

(e) Heifers of estimated stages of pregnancy (from Schmid [16]).

Stage of pregnancy (mths.)	1	2	3	6	8	10*
No. of animals	2	2	2	1	1	1
Sum weight of two ovaries (gm.)	—	14·3	13·9	—	10·4	22·5
Ratio $\dfrac{\text{Ovary with corpus luteum}}{\text{Ovary without corpus luteum}}$	—	198	163	—	174	137
Size of corpus luteum (cm.)	2·0	2·0	1·8	2·3	2·1	1·9

* A cow.

heifers of the same type and age that there is in pregnancy a slight increase in the weight of the ovary containing the corpus luteum in proportion to the other ovary above that which occurs at mid-period of the cycle. Such increase can be accounted for by slight increase in the diameter of the corpus luteum as has been shown above.

As can be seen in Table XXV from our own results (*a*) and (*b*) and from those of Küpfer (89) (*d*), Schmid (16) (*e*), and Bergmann (257) (*c*) there is no alteration in the ratio $\dfrac{\text{weight of ovary with corpus luteum}}{\text{weight of ovary without corpus luteum}}$ due to the period of pregnancy, thus proving indirectly that the corpus luteum maintains its size throughout pregnancy.

Table XXV also shows that there is no constant change in the sum weight of both ovaries during pregnancy which negatives the idea that there is any hypertrophy of the stroma or interstitial cells of the ovary (if such exist) in the cow during pregnancy.

The ratio $\dfrac{\text{weight of ovary with corpus luteum}}{\text{weight of ovary without corpus luteum}}$ varies with the age of the animal as will be seen from Küpfer's results (Table XXV (*d*)), being greater in heifers than in cows owing to the accumulation of ovarian stroma in the latter (see p. 47). Variations that exist in the ratio of one ovary to the other during pregnancy (see Table XXV) depend on the total weight of the ovaries, the larger the total weight (owing to ovarian stroma growth) the smaller the ratio. This cause will explain the variations which occur in the ratio of the two ovaries during pregnancy in different animals and it can be said that the cyclic alteration in ratio between the two ovaries due to the formation and atrophy of the corpus luteum (see Table IX) and the ripening of follicles is absent during pregnancy.

Corner (107) found in the sow that, in the cycle, the time of commencing involution of the corpus luteum (15th–19th day) normally begins a few days after the period at which the foetal membranes have become attached to the uterus in the pregnant animal (10th–15th day). Much the same time relations occur in heifers; the foetal membranes of one killed at the end of the first month of pregnancy had not yet become attached to the cotyledons by foetal projections. In the closely related sheep in which heat occurs every fortnight or three weeks the blastodermic vesicles remain free till the 17th day (Assheton (258)). Meyer (259) concluded that the life of the granulosa cells or corpus luteum under various conditions synchronised with the life of the ovum which they nourished.

Hammond (256) found in the rabbit that, by removing the foetuses and leaving the foetal membranes attached to the uterus, the cause of the persistence of the corpus luteum originated in the foetus. Loeb (47) however states that the production of deciduomata in the guinea-pig prolongs the cycle and he concludes that the production of these bodies in the uterus prolongs the life of the corpus luteum, but he made no direct measurements of the size of the corpus luteum under these conditions. He also found in a case of extra-uterine pregnancy that the foetus did not prolong the life of the corpus luteum. Long and Evans (32) in the rat found no delay in ovulation following insertion of threads into the uterus, although no decidua were produced under these conditions; they found however that when the life of the corpus luteum is prolonged, as in lactation, deciduomata can be produced by this method and Loeb's conclusions are open to the criticism that variations in length of the life of the corpus luteum affect the formation of deciduomata rather than that the deciduomata cause prolongation of the life of the corpus luteum. The same criticism applies to Zschokke's (40) statement that an irritation of the cow's uterine mucosa following infection (such as occurs in contagious abortion and granular vaginitis) causes the corpus luteum to persist. While this supplies a very plausible argument as to the cause of persistent corpora lutea (and they are said to be frequent after contagious abortion) we do not consider that it has yet been proved that foreign bodies in the uterus will cause persistent corpora lutea. Krainz (260) found in bitches that long-retained foreign bodies in the uterus do not hinder the onset of a new heat period, although in this animal true deciduomata are not found but only greater glandular growth occurs.

Whether the action of the foetus on the life of the corpus luteum is one of direct stimulation by means of internal secretions (which is rendered doubtful by Loeb's case of extra-uterine pregnancy) or is due to the continual utilisation from the uterus by the foetus of either the products of, or substances similar to, those which probably cause the death of the corpus luteum (p. 81) is at present unknown. The latter view obtains some support from the fact that large quantities of lipoid are found in the placenta (Bienenfeld (261)). In birds large quantities of lipoids are stored in the yolk but in mammals the bulk would have to be supplied by the placenta. Smith (262) has shown that in crustaceans the lutein is formed in the liver and transferred to the ovary. Loeb's and other investigators' experiments indicate that continuous new stimulation by cutting or foreign bodies is necessary for the development of deciduomata and it may be that the accumulation of the products of development hinders further growth unless it be removed by a living foetus or destroyed by a wound.

Ewart[263] found in the mare that gestation is most likely to become interrupted and abortion to follow between the 6th and 9th weeks of pregnancy during which time the embryo loses its primitive attachment to the uterus before acquiring its more permanent connections. The mere freedom of the membranes will not explain abortion under these circumstances, for a similar free unattached state occurs before the primitive attachments are made and abortion does not occur then. It is more probable that the life of the corpus luteum is affected at these times by one or other of the means indicated above.

It has frequently been stated that the corpus luteum is not necessary for foetal life and the continuance of pregnancy during the latter half of gestation (Marshall and Jolly[264], Kleinhaus and Schenk[265]); others have gone so far as to say that the corpus luteum undergoes involution during the second half of pregnancy. That the latter is untrue as far as size and structure goes is shown in Table XXIII and by Bergmann[257] in the cow, by Hammond[256] in the rabbit and by Corner[266] in the sow. Hess[41] states that squeezing out the corpus luteum from the ovary is the simplest way of causing artificial abortion in cows and that abortion occurs within 24–48 hours after the operation. Schmaltz[1] also states that removal of the corpus luteum in the pregnant cow invariably causes abortion and quotes Zieger to the effect that this occurs 48–54 hours after the operation (cf. time of heat after removal of corpus luteum, p. 15). We have no experimental evidence on this point in the cow but in the rabbit removal of the ovaries in the latter half of pregnancy has always been attended by abortion (Hammond and Marshall[24]) although the time at which abortion occurs after removal appears to increase with the stage of pregnancy. This confirms previous work on the rabbit by Blair Bell and Hick[267]. Weymeersch[62] found that removal of the corpus luteum in rabbits caused foetal atrophy and absorption during the first half of pregnancy but caused abortion during the second half. Fraenkel[42] found in the rabbit that foetal atrophy occurred after the removal of the corpora lutea, but that the number of the corpora lutea could be reduced to less than one-half of the number of the foetuses before development was interrupted. Küpfer[89] found in the cow that in cases of twin pregnancies the size of each corpus luteum was slightly smaller than in the corresponding period of a single pregnancy and that when one foetus degenerates both corpora lutea develop as though two normal embryos were present.

Follicular degeneration. The persistence of the corpus luteum throughout pregnancy will explain why the rhythm of the reproductive organs is absent during gestation since it has been shown (p. 15) that the life of this body controls the rhythm during the cycle. Fellner (268) believes that in woman there is a rhythm in the ovaries during pregnancy but he based his conclusions on blood pressure readings and not on direct observation of the ovaries themselves.

During pregnancy follicles may attain their full size (excluding the pre-ovulation increase) but in those heifers killed during the 6th and 7th months no large follicles were present (see Table XXIII). The occurrence of large follicles in some cases shows that the presence of a corpus luteum in the ovary does not necessarily prevent a Graafian follicle from increasing in size, as is also seen by the gradual increase in size of the follicle during the cycle (p. 41). The presence of an active corpus luteum does however prevent the pro-oestrous increase in size and also the subsequent ovulation. These results are more or less in accordance with those obtained by Loeb (269) in the guinea-pig.

Whether the follicle which has attained large size (and would normally rupture at the next heat period had not pregnancy occurred) persists throughout pregnancy or whether it atrophies and is followed by the development of others has not been determined; examination per rectum of ovaries in which only one follicle is shed at each heat period (such as the cow) at intervals during the period of pregnancy would decide this.

Follicular atrophy is however greatly increased during pregnancy. A comparison of the frequency of atrophic follicles in the ovaries of non-pregnant and pregnant heifers shows that there is a great increase of atrophy during pregnancy and few ovaries of pregnant cows or heifers (more especially in the later stages of pregnancy) have been seen without large numbers of atrophic follicles.

Küpfer (89) found that the number of follicles in the ovary visible to the naked eye decreased towards the middle and end of pregnancy. The extent to which atrophy occurs no doubt depends on the food content of the special substances required for follicular growth, the availability of which in pregnancy is reduced by the requirements of the corpus luteum. That feeding affects atresia of the follicles has been shown by Loeb (33) in the guinea-pig and similar changes occur during lactation on a moderate diet in rabbits (Hammond and Marshall (24)).

PLATE XXI

1. Atrophic follicle—early stage.

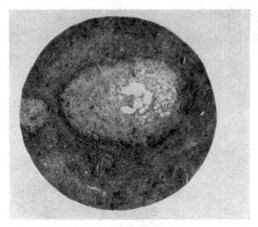

2. Atrophic follicle—mid-stage.

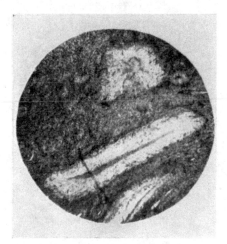

3. Atrophic follicle—late stage.

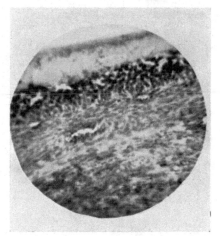

4. Wall of small follicle.

Follicular atrophy in the cow has been described and figured by several investigators. Delestre (100) found that the granulosa disappears and connective tissue ingrowths from the theca occur; sometimes the granulosa cells become detached *en bloc* and are seen floating in the liquor folliculi; this we can confirm.

Schmaltz (83) also describes and figures stages in the degeneration process with ingrowing connective tissue. Marshall and Peel (53) have also figured atrophic follicles from very fat heifers which show in the centre remains of the granulosa surrounded by a mass of ingrowing connective tissue. Simon (109) found atrophic follicles in cow's ovaries and describes them as containing a centre of degenerate granulosa cells and a shrivelled flattened ovum surrounded by a rosette of clear cell-free connective tissue; he also figures various stages of atrophy. A similar clear layer between the degenerate granulosa and ingrowing stroma has been figured by van Beck (107) in the degenerate follicles of calves (late foetal or a few days old).

Our own observations show that during pregnancy many of the smaller follicles degenerate before becoming mature. The appearance of degeneration is first seen in the granulosa which frequently becomes detached from the theca and tends to collect in a mass in the middle of the follicle while the liquor folliculi is absorbed and the cavity is filled in by a growth of connective tissue cells from the theca. Occasionally a few of the granulosa cells escape the general degeneration and hypertrophy, forming small luteal cells. Successive stages of follicular degeneration are shown in Plate XXI, 1 to 3.

Another type of follicular atrophy, distinct from the above, and associated with bleeding into the follicular cavity also occurs in the cow (see p. 183, Cow A 4). Käppeli (105) has described and figured this type in the ovaries of calves of the Russian Steppe breed at 2 months old and in a Simmenthal calf of 5 weeks old; he states that degeneration was caused by chromatolysis of the granulosa and blood entered the follicle from the theca interna. A similar type of atrophy occurs quite commonly in the rabbit (Hammond and Marshall (24)).

(b) *The Uterus and its Contents.*

The economic importance of the changes occurring in the pregnant uterus of the cow, which are described below, lies very largely in the comparison of these normal changes with those that occur in cases of contagious abortion and which should present evidence as to the way in which parturition or abortion is brought about. In addition the

H 9

changes throw light on the food requirements of pregnant animals and the nutrition of the foetus, with the possibility also that some of the changes described may affect the growth of the udder.

(1) *The Size and Proportion of the Parts.*

The increase in the size of the uterus and its contents during pregnancy is shown in Table XXVI (*a*) and in Diagram II, from which it will be seen that the rate of gain is very rapid and increases in amount and in actual, although not in relative, velocity as pregnancy advances up to the 8th month. During this time the weight is increased a hundredfold, from 320 gm. at the end of the 1st month to 32040 gm. at the end of the 8th (4 weeks) month of pregnancy.

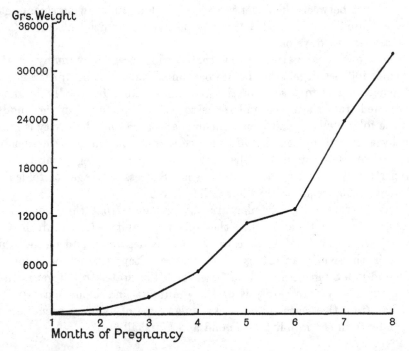

Diagram II. Increase in weight of uterus and its contents during pregnancy.

The composition of this increase in weight is seen in Table XXVI (*a*). Unfortunately in the bulk of the material the foetal membranes were not separated from the cotyledons of the uterus when in a fresh condition, but this was done after fixing in formalin; in this condition the foetal placenta did not pull away from the cotyledons easily so that in our results the weight of the uterus is relatively too high and that of the foetal membranes too low. The figures for the different months are however comparable except those of P 9 in the 8th month, when the membranes were separated in a fresh condition and in which, as compared with the others, the membranes are too high and the uterus too low.

Similar weights for the components of the pregnant cow's uterus have been given by Colin (270) and Franck-Albrecht (244, p. 97), but in most cases the exact period of pregnancy was unknown.

The changes occurring in the pregnant uterus may be viewed in two ways: (1) the actual increase in weight of the different parts and (2) the relative growth of the parts or percentage which each forms of the whole at different stages.

(1) Diagram III shows that the actual increase in weight is due to different parts in the various stages; from first to last however the increase is mainly due to the foetus, then to the foetal fluids, and to a smaller extent to the growth of the uterus and foetal membranes.

The foetus however, as will be seen from this diagram, plays but a small part in the increased weight at first but takes a larger and larger share in the increase until towards the end the bulk of the increase in weight is due to foetal growth. While up to the 3rd month the foetus forms the smallest part of the whole yet at the 8th month it weighs more than all the rest of the parts together. The growth curve given corresponds very closely with that shown by Vignes (188) for man.

Our own material has been limited so that no opportunity has been obtained for finding the variation in weight of the foetus at any one age. Eckles (271) and Haigh, Moulton and Trowbridge (272), have shown that the birth weight of calves varies with the breed as well as nutrition and other causes; foetal development will no doubt vary in the same way. The weights of cow embryos given by Rörik (273) of ages similar to ours are shown in Table XXVI (c), but the variation is so wide that we are inclined to believe that he was mistaken in the age of some of the cases, in particular those at 2 and 6 months.

Buchem (274) gives the weight of a number of cow embryos in different weeks and months of development and the variation in weight at any one age is considerable, the coefficient of variability being about 0·4 during the first two months and afterwards in the region of 0·1; the averages for his foetuses at ages corresponding to our own are given in Table XXVI (c). Bergmann (257) also found considerable variation at

9–2

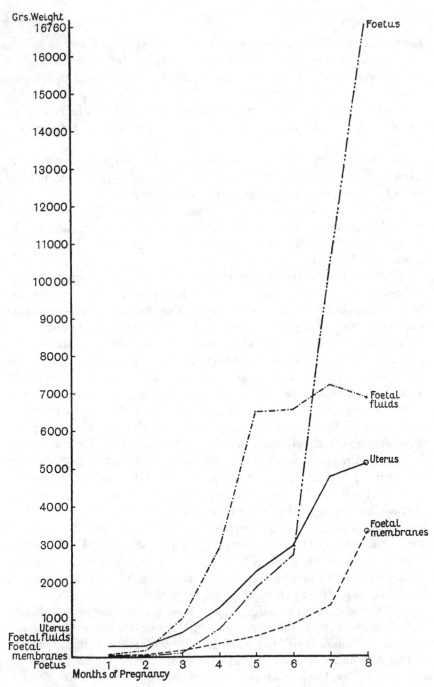

Diagram III. Composition of uterus and contents during pregnancy.
Actual weights—in grammes.

any one age in the weight of calf foetuses of the spotted Niederungs breed, and he gives the age of the foetus to the day; the weights of his foetuses which correspond to ours in age are shown in Table XXVI (c) and the variation in weight of three foetuses—470 gm., 650 gm. and 1270 gm.—each 122 days old is remarkable. Ewart[246] found in the mare a considerable variation in development during the 3rd week of pregnancy from embryos previously described by Martin and also Bonnet at the same age, and since the mare remains on heat a long time he attributed it to difference in the time of service during heat. However, Kirkham[275] has shown that in the mouse the development of the embryo may be retarded by lactation of the mother; our results are not affected by this cause as all were from heifers pregnant for the first time; the variation found by Buchem and Bergmann is however open to this possibility.

Meyer[276] found in women that there was a great difference in length and weight even in twins and Hammond[277] found this to be the case in the embryos in litters of pigs and rabbits, so that differences in weight and length of the foetus at any age are probably due to inherited characters and nutritional causes in most cases and not necessarily to variations in time of gestation as might be supposed.

The age of a foetus can be estimated from its weight or its length (Table XXIX), but it has been found by most observers that there is a greater variation in weight than in length, and Bonnet has pointed out in the sheep the frequent lack of correlation between the weight and length growth of the foetus. We believe that length growth is the best criterion of age for we have found in studying the growth of the sheep[278] that deficient nutrition (such as may occur during foetal development) affects weight growth (or muscle and fat development) before it affects length growth (or bone development). Haigh, Moulton and Trowbridge[272] found in new born calves that the heavier ones showed a higher proportion of flesh and lower proportion of skeleton than the lighter calves of the same breed. The various factors which affect foetal growth as determined from birth weight have been investigated by Hammond[279] in the rabbit and by Eckles[271] in the cow. Hart, Steenbock and Humphrey[280] found that rations restricted either to the wheat or oat plant affected the proper development of the foetus in the cow. Eckles[281] concluded that a cow can produce a foetus on the same ration that will only maintain her when barren; if this is true the pregnant animal must be able to utilise her food to better advantage or the energy requirements of a number of oestrous cycles is equal to that metabolised by pregnancy, and the latter as Eckles shows from chemical analysis of the foetus, placenta and foetal fluids just before parturition is small, for their percentage of water is large; Haigh, Moulton and Trowbridge (272) have investigated the chemical composition of the cow foetus at different ages. Since very fat cows frequently produce undersized calves, and it has been shown above (p. 82) that the fat clogs the nutrition of the ovary and affects its internal secretions, it is probable that the nature of the food supply is a more important factor than the quantity in foetal nutrition.

The foetal membranes increase regularly in actual weight; during the first three months their growth exceeds that of the foetus but afterwards lags behind its growth. Bergmann (257), by counting the attachments of the membranes to the cotyledons, concluded that no increase in the number of these took place after the 3rd month; it is probable that after this period (3rd month) growth of the cotyledons in size occurs as well as growth and thickening of the membranes between them, but no further extension of the membranes takes place towards the apex of the horn of the non-pregnant side.

Bergmann also gives data as to the weight of the uterus and foetal membranes together during different stages of pregnancy but he did not separate them. His figures for animals similar in age to our own are shown in Table XXVI (c). While they more or less agree with ours in the later stage of pregnancy they are much heavier in the early stages, owing to the fact that he included cows in which the uterus had been developed by a previous pregnancy while ours were first calf heifers. This difference in weight may possibly explain why the offspring of first calf heifers is usually smaller than that of mature cows. Our results show that in animals previously virgins the uterus itself does not begin to increase much in weight until after the 2nd month of pregnancy, but after this time it makes steady and regular increases up to at any rate the 8th month of pregnancy. There is nothing in the shape of the growth curve of the uterus itself (see Diagram III) which would suggest the formation of a myometrial gland at or about the 5th month such as Ancel and Bouin (282) have suggested to account for the glandular growth stage of the mammary gland which occurs in the second half of pregnancy in the rabbit.

The foetal fluids increase very rapidly during the early stages of pregnancy up to the 5th month and after this time their amount remains nearly constant. Bergmann (257) gives weights of both the allantoic and amniotic fluids in the different stages of pregnancy, and his data (see

Table XXVI. *Increase in weight and percentage composition of the uterus and its contents during pregnancy.*

(a) *Actual weights (gm.).*

Month (4 weeks) of pregnancy	1	2	3	4	5	6	7	8
No. of heifer—pregnant for first time	P 2*	P 4	P 3	P 1	P 5	P 7	P 6	P 9†
Uterus	303·0	300	650	1310	2280	2940	4770	5120
Foetal membranes	2·1	50	150	350	520	850	1370	3320
Foetal fluids	15·0 (?)	181	1053	2920	6510	6540	7220	6840
Foetus	0·1	9·2	87	720	1830	2700	10380	16760
Total uterus and its contents	320·2	540·2	1940	5300	11140	13030	23740	32040

(b) *Percentage of total uterus and its contents.*

Uterus	94·6	55·5	33·5	24·7	20·5	22·6	20·1	16·0
Foetal membranes	0·7	9·3	7·7	6·6	4·7	6·5	5·8	10·4
Foetal fluids	4·67	33·5	54·3	55·1	58·4	50·2	30·4	21·3
Foetus	0·03	1·7	4·5	13·6	16·4	20·7	43·7	52·3

(c) *Actual weights (gm.) from results of other investigators.*

Buchem (274)	Foetus	0·27	8·3	109	599	1712	5056	9450	16075
Bergmann (257)	Foetus	—	7·0	185	470‡ / 650 / 1270	2650	6300	9600	19300
	Uterus and foetal membranes	—	456	1190	2050‡ / 2300 / 2990	4150	8450	7400	8100
	Foetal fluids	—	402	1075	1960‡ / 3150 / 3790	7600	3450	6600	9100
	Allantoic fluid	—	365	410	500‡ / 1100 / 1200	3300	4200	3450	5500
	Amniotic fluid	—	37	665	1460‡ / 2050 / 2590	4300	4250	3150	3600
Rörik (273)	Foetus	—	500	77	480	6500	325	23000	20000
	Uterus	—	1000	615	1356	3600	1830	3800	6250
	Foetal fluids	—	1500	718	2020	—	1675	—	—
	Foetal membranes	—	81	30	144	2900	170	5200	4750

* Uterus and contents not weighed whole.

† Uterine cotyledons separated from foetal membranes in fresh condition.

‡ Three separate cases at same age.

Table XXVI (c)) at similar ages to our own show that there is some but not much increase in the amount of fluids after the 5th month, the increase if any taking place in the allantoic fluid while the amniotic fluid shows a decrease after this time. Whether the cessation of the increase in foetal fluids at the 5th month of pregnancy is connected in any way with the definite changes which occur in the udder at this time (see p. 168) it is not possible to say without experimental evidence; but it is just possible that the foetal fluids may require ovarian secretion for their production and that after the 5th month this becomes available for the mammary gland, or an actual absorption of foetal fluids into the maternal circulation might supply a cause.

Bergmann (257) found that in cows with twin pregnancies (at the 2nd and 5th month) the weight of the foetal fluids was very much above normal, the difference in the fluids being much more marked than in the case of the other parts of the pregnant uterus. The amount of the foetal fluids is said to be greatly increased in cows served by a Bison bull (Boyd (283)) and frequently leads to abortion; whether the secretion is a result of the increased vigour of the cross or is due to pathological development is not known, but there is no similar trouble in the second and third crosses. Crew (284) has pointed out that hydramnios also occurs about the 3rd–4th month of pregnancy in association with the formation of bulldog calves in the Dexter breed of cattle.

Bergmann (257) found no changes in the length of the Fallopian tubes during pregnancy.

It will be seen from the above account that the actual increase in size of the pregnant uterus and its contents is due in the early stages of pregnancy mainly to the growth of the foetal membranes and fluids and in the later stages of gestation to the growth of the foetus and to a lesser extent of the uterus itself. It is remarkable that there is very little increase in actual weight of the foetal fluids after the 5th month.

(2) The relative development of the parts or proportion which each bears to the total weight of the pregnant uterus and its contents is shown in Table XXVI (*b*) and in Diagram IV. This diagram shows clearly the relation of the different parts and how they change during the course of pregnancy. The principle of the chemical law of mass action applies to this relationship; the development of the foetus in the latter part of pregnancy is seen to be outstripping the provision of the nutritional parts—the uterus and foetal membranes (a fact which has been emphasised by Rörik (273)). This factor may be concerned in the cause of parturition.

Diagram IV also presents a picture of the physical conditions which exist at the different periods of pregnancy. The development of the foetal fluids in the early stages expands the uterine cavity in preparation for the developing foetus which in turn develops at the expense (in weight) of the foetal fluids rather than by much further enlargement of the uterus itself. The large amount of foetal fluids present in the early stages of pregnancy in cases of twins (see Bergmann (257)) is provision for the future accommodation of the foetuses. Whether the variability in weight of the foetus and foetal fluids at any particular age is the result of this relationship must be left for future investigation to decide. It would appear that in the bulldog calves of the Dexter breed a dwarfing of the foetus results in a hypersecretion of foetal fluids or *vice versa*.

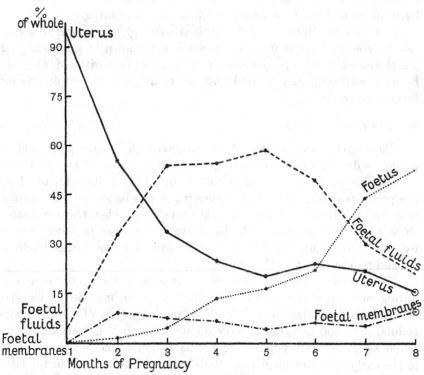

Diagram IV. Composition of uterus and contents during pregnancy as percentage of whole.

The relation of the time at which the proportion of the foetus itself exceeds that of the foetal fluids (7 months) is significant in that it is usually about this time that abortion most frequently occurs in cases of infection with *Bacillus abortus*. Whether or no infection before this time (about the 7th month) causes foetal absorption (as happens in the first half of pregnancy in rabbits when the ovaries have been removed) is not known, but the prevalence of sterility in infected herds, with cows going some months without showing signs of heat and then coming on heat again (see Williams (285)), would suggest that this is so. Whether the proportion between foetus and foetal fluids changes in rabbits (and the percentage of the former exceeds that of the latter) at a time when the removal of the ovaries leads to abortion instead of absorption *in situ* is also unknown, but it appears probable. Williams[286] states that when abortion does occur before the foetus is about 30 cm. long it is expelled enclosed in the foetal membranes but after this period it is born naked and the membranes expelled as an afterbirth.

How the relationship of the various parts of the pregnant uterus and its contents varies in other species is not known in any detail nor has the effect of these proportions on the cause of parturition or abortion been investigated, and it would appear to afford a profitable line for future investigation.

(2) *The Uterus.*

The increase in the size of the uterus itself during pregnancy is probably due to three main causes: (*a*) the influence of the persisting corpus luteum in raising uterine nutrition, (*b*) the irritation caused by the presence of the foetal membranes which can be seen in the proportionally large increase in the size of the cotyledons when the membranes come into direct contact with the uterus, and (*c*) the pressure exerted by the foetal fluids at first and afterwards by the foetus itself in expanding the cavity of the uterus.

The muscular coat. The increase in the muscular coats of the uterus during pregnancy is brought about mainly by an increase in the size of the muscular fibres. As will be seen from Table XXVII, in the longitudinal muscular coat the average number of muscle nulcei in a microscopic square is 22 in virgin heifers of $2\frac{1}{2}$ years old and is reduced to 11 in the early stages of pregnancy (2nd–3rd months) and to 4 in the later stages of pregnancy (7th–8th months). That is, the cytoplasm of the plain muscle fibre has increased considerably, showing that the enlargement of the muscle itself is due to increase in the size of the muscle fibres rather than to any great increase in their number. Evidence as to the formation of new muscle fibres is difficult to obtain but in microscopic sections it is not so evident as increase in the size of the fibre; the latter can be observed easily and the cytoplasmic increase is so marked that towards the end of pregnancy it appears to be clear and vacuolated by the usual methods of staining (Fe Haem. and Eosin).

Wall[206] found that the muscular coat of the pregnant uterus is not rich in nuclei but that the virgin uterine muscle is; he states that the involuted uterine muscle contains even more nuclei than the virgin.

Rab[287] states that the greatest muscular hypertrophy occurs in the 5th–6th months. Schmaltz[1] quotes several investigations which have shown that the muscle fibre increases in length during pregnancy even more than it increases in thickness, from 115μ before pregnancy to 700–850μ after the 20th week of gestation. Although the muscle fibres of the uterus hypertrophy in pregnancy the thickness of the wall of the uterus decreases owing to the extension which occurs. Zieger[110] has shown by measurements that this is so and that the thickness of the pregnant side in which the extension is greatest decreases more (8 mm. in virgin to 1·5 mm. at 8th month of pregnancy) than that of the non-pregnant side (8 mm. to 3 mm. at 8th month).

Table XXVII. *Average number of nuclei in a given area in cross-sections of the longitudinal muscular coat of the uterus in different reproductive states.*

Reproductive state			No. of heifer	Number of muscle nuclei in a micrometer square (av. of 10 squares)	Average
Oestrous cycle		48 hrs. after heat	C 4	19·1	
		72 ,,	C 6	29·9	22·2
		8 days after heat	C 2	17·5	
Pregnancy	Early	2 months	P 4	14·4	
		3 ,,	P 3	8·4	11·4
	Late	7 ,,	P 6	4·7	
		8 ,,	P 9	3·1	3·9

Rab[287] found that during pregnancy the longitudinal muscular coat increased in thickness mainly in two bands one along the large curvature and one along the small curvature and that this thickness was most marked at the 5th–6th month and after; Zieger[110] has confirmed his observations. We however have carefully examined the fixed muscle at all stages of pregnancy but could find no differences in thickness of the muscle in any special area of the uterus.

Not only is there a change in the muscle of the muscular coats of the uterus during pregnancy but an alteration in the character of the connective tissue binding also occurs. The connective tissue cells become enlarged and swollen in a similar way to those of the mucosa, which give rise to decidual cells, but to a less extent. This is of interest in connection with the myometrial gland described by Ancel and Bouin[288] in the rabbit, since they associated the occurrence of this gland with the phase of secretory activity of the mammary glands such as begins in heifers at about the 5th month of pregnancy. The appearance of the connective tissue in the vascular area between the longitudinal and circular muscle coats where this gland is supposed to occur has therefore been carefully examined at different stages of pregnancy. No trace of the myometrial gland has been found but the connective tissue cells become enlarged and vacuolated and in the mucosa hyaline deposits occur in and between these cells in the early stages of pregnancy as a result probably of raised nutritional conditions existing.

Fraenkel(289) also failed to find the myometrial gland in the cow and other animals although he describes it in the rabbit. Hammond(256) occasionally observed it in the rabbit's uterus, but its occurrence is too irregular and uncertain for it to have the important functions which the discoverers attributed to it.

The vascular layer between the circular and longitudinal muscular coats which is rich in connective tissue cells increases greatly during pregnancy and is much more marked in the uteri of cows which have borne calves than in animals which have never become pregnant. This is due to the fact that the connective tissue and vascular growth which takes place in pregnancy is never completely reabsorbed afterwards.

The actual size of the arteries in the vascular zone in the different stages of pregnancy in the cow have been determined by Hilty(205). He found that while the maximum size of the arteries in the vascular zone at the 6th week of pregnancy was, lumen 80μ and walls 130μ—by the 15th week it had increased to 360μ and 460μ respectively, and in the 26th week was 620μ and 380μ respectively. Post-partum the arteries become smaller but mainly in the lumen, for the walls are thickened by connective tissue growths; at 29 days post-partum the lumen measured 112μ and the walls 256μ. These changes are similar to those which occur in the blood vessels of the cotyledon (see p. 92).

The changes in structure between the virgin and pregnant uterus are described by Bruyn-Ouboter(290) in cats, dogs and rabbits as consisting of a thickening and broadening of all layers of the uterus but particularly of increase of blood vessels and broadening of the stratum vasculosum.

The intercotyledonary area. The area situated between the cotyledons is also considerably modified by pregnancy. Beginning at the 2nd–3rd month the whole intercotyledonary area is thrown up into a number of folds or wrinkles in hardened specimens and these increase in size as pregnancy advances. Reference to the histological description above (p. 87) will show that this area is occupied chiefly by glands and these during pregnancy secrete the uterine milk which nourishes the foetus both before and after its attachment to the uterus. The secretion—uterine milk—has been analysed by Gamgee[291] who found it contained 10·5 per cent. protein, 1·25 per cent. fat as well as salts. Kolster[207] by histological methods identified fat in the secretion from the uterine glands in the cow. There is difference of opinion as to the origin of uterine milk, for whereas Haller and others say that it originates in the uterine glands, Turner, Ercolani and others[292] state that it can be squeezed out of the crypts of the cotyledons as a pinkish white fluid. The cotyledonary crypts (see p. 148) have no maternal epithelium and are not secretory in the true sense of the word; this area too is quite free from the openings of any uterine glands. A secretion however can be obtained by squeezing a cotyledon which has been separated from the foetal membranes. This fluid is lymph—much the same as chyle—with numerous leucocytes and nutritive substances and a fair number of red blood corpuscles which exude from the surface of the cotyledon wounded by the ingrowing foetal tissues. It also probably contains cells of the foetal trophoblast which are detached when the foetal membranes are pulled out. This fluid is not formed by the cotyledons before attachment takes place as the crypts do not exist as formed elements before the ingrowth of the foetal tissues. Kolster[293] has shown that in the pregnant mare the uterine mucosa is very rich in lymphatic vessels and also that in the cow[207] the lymphatics develop greatly in the mucosa of the pregnant uterus; he found globules of fat in the superficial layers of the maternal projections of the cotyledons as well as in the uterine glands.

Wall[206] describes the walls of the crypts in the cow as consisting of fine connective tissue with fibroblasts and emigrating scattered lymphocytes and polymorphonuclear leucocytes. The uterine glands which open only on the intercotyledonary area also produce a secretion in both early and late stages of pregnancy; this secretion, which contains protein fat and leucocytes, is necessary for the nutrition of the blastocyst before the cotyledonary attachments are formed as well as supplementing their function afterwards. Schauder[294] has recently published a review of the literature of the morphology of uterine nutrition in the domestic animals.

The appearance of the uterine glands varies considerably during pregnancy. At the end of the first month the glands lying next the surface have a fairly large lumen as compared with that which occurs during the cycle (p. 89), but the deeper parts of the glands next the muscular coat are still small (Plate XIII, 4).

At the 2nd month the glands have increased in size and their lumen is larger but the deeper glands are still small; detritus masses can be seen in the lumen of the glands near the surface. At 3 months all the glands both superficial and deep show a large lumen and signs of marked lateral pressure may now be observed, for they are no longer circular in outline but their long axis lies parallel to the surface of the uterus. They have a large lumen which contains detritus—the products of their secretion; the epithelium is cubical and regular and gives the appearance of true secretion occurring rather than cell disintegration.

At the 4th month very similar conditions were observed but the glands did not appear quite so compressed as in the last (Plate XIII, 5).

If Plate XIII, 5 at the 4th month be compared with Plate XIII, 4 at the 1st month of pregnancy it will be seen that the number of glands passed through from muscle to epithelium is much less in the former than in the latter; this is to be explained by the increase in area of the uterine surface and the extension of its coat and not by diminution in the extent of the glands which remains the same as heretofore. This process of thinning of the glands by extension of the uterine coat goes on during the 5th and 6th month so that only one or two glands appear in the line between muscle and epithelium, this distance being much less than at 1 month. Nevertheless the glands that exist have increased greatly in diameter and have very large lumina which contain the products of secretion.

In the 7th and 8th months the glands are enormous in size and with very large lumina; the pressure exerted on them by the foetal fluids has caused them to be compressed and it is their long axis, which lies parallel to the uterine surface, which is increased most (Plate XIII, 6); they are filled with the products of secretion. In the 8th month of pregnancy the openings of the glands are so large that they can be observed by the naked eye as little pits on the surface of the mucosa.

The relative sizes of the glands at the different stages of pregnancy are shown in Table XXVIII (*a*). Kolster [207] has commented on the difference in size of the lumen of the superficial and deep glands in the 5th month of pregnancy in the cow. Hilty [205], who made measurement of the glands during pregnancy in the cow, also found increase in size both of the lumen and epithelium, but his figures (Table XXVIII (*b*)) do not show much increase in size after the 3rd month and on this point we disagree with him; his figures also show that there is a marked distinction between deep and superficial glands during the latter part of pregnancy but in our experience there are at this time only about two layers of glands of about the same size. Hilty showed also that after

Table XXVIII. *Relative sizes of uterine glands in different stages of pregnancy.*

(a) *Diameter of cross-section (micrometer squares).*

Month of pregnancy	No. of heifer	Parallel to uterine surface	At right angles to uterine surface	Average diameter
1	P 2	1	1	1
2	P 4	5	5	5
3	P 3	1	7	4
4	P 1	2½	5½	4
5	P 5	3	5	4
6	P 7	4½	10	7
7	P 6	5½	18	12
8	P 9	4	46	25

(b) *Diameter in μ. (From Hilty [205].)*

	Lumen		Epithelium	
Reproductive state	Surface glands	Deep glands	Surface glands	Deep glands
Non-pregnant heifer 17 months old	22	8	12	12
Pregnant 6 weeks	40	6	16	8
9 „	80	32	36	16
13 „	160	56	28	20
15 „	160	44	28	16
20 „	160	46	24	14
26 „	160	40	22	16
Post partum 3 days	144	50	20	16
15 „	80	44	10	9
21 „	45	14	16	15

parturition the glands involuted again quickly. Wall [206] found that the average diameter of the uterine gland tubules in heifers was 17μ; during pregnancy (7–9 months) it rose to 150μ; 3 days post-partum it was 50μ and in the involuted uterus 17μ. It will be seen that large size of the uterine glands is associated with the presence of a corpus luteum in the ovary and that while it is present they remain large and when it atrophies they become small again.

The conclusion arrived at from a study of the uterine glands during pregnancy is that they do not increase much in length or number but increase enormously in diameter and activity, the lumen becoming large and the epithelium active; it would appear that secretion from them takes place in increasing quantities with the progress of pregnancy and that secretion is not caused by breakdown of the cells but by secretion proper.

We have observed doubled-in folds of the epithelial cells in the glands (see Plate XIII, 6) similar to those described by Kolster [207] in the mare and cow but do not agree with him that they are formed for secretory purposes, but rather that they are caused by pressure of the uterine contents doubling the winding glands upon themselves and leading to squashing of the epithelial cells at the corners of the folds. Wrinkles of epithelium form in the gland tubes also as a result of collapse of the tissue after the pressure is removed and it is fixed. In the cow Kolster found that these invaginates were more frequent in the deep parts of the glands than near the surface and this we believe is due to the fact that the surface part runs more vertically and is less convoluted than the deeper part.

The connective tissue in the deeper parts of the mucosa surrounding the glands contains masses of hyaline matter in the 1st month of pregnancy, in very much the same way as it exists at the latter end of the cycle (see p. 89); the connective tissue cells also appear to be swollen (Plate XIII, 4).

A little of this hyaline matter is still to be seen at the 2nd month of pregnancy but afterwards it disappears, probably being utilised in growth or by the foetal tissues. Jenkinson(295) states that he has found glycogen both in the uterine and glandular epithelium of the pregnant cow and in the sub-epithelial connective tissue of the sheep's uterus, but we believe that the substance seen in heifers is not glycogen but rather material similar to the Whartonian jelly of embryonic connective tissue. With advancing pregnancy the connective tissue in this area becomes looser and looser so that in the later stages of pregnancy in fixed material the outer part of the mucosa (sub-epithelial layer) can be easily stripped away from the muscle. This loose character of the connective tissue in pregnancy Kolster(207) believes is due to the formation of large and numerous lymphatics; he however describes these as being rather less frequent in the later stages of pregnancy. The formation of large lymphatic vessels is however not the only cause, for stretching of the uterine surface tends no doubt to thin out the tissue unless special growth occurs and while this special growth does occur in the sub-epithelial layer and in the muscle much less takes place in the glandular layer.

It has previously been noted (see p. 87) that in the uteri of virgin heifers there is a thin layer of dense sub-epithelial connective tissue (see Plate XIII, 1 to 3) similar to that of which the cotyledons are composed. At the end of the 1st month of pregnancy this layer is very similar to that of the virgin condition and contains free red blood corpuscles and distended capillaries (Plate XIII, 4).

As pregnancy goes on this layer increases in thickness in a similar but to much less degree than that of the cotyledons and forms a thickened gland-free (except where the mouths of the glands penetrate it) dense connective tissue mass in which blood capillaries are numerous (see Plate XIII, 5 at 4 months and Plate XIII, 6 at 7 months). It is from this layer that adventitious cotyledons are formed (see p. 155) and as a rule it is thicker in those parts of the surface that are pressed closely against the foetal membranes, *i.e.* the upper parts of the wrinkles on the uterine surface. Kolster(207) has figured the increase in thickness of this layer during pregnancy in the cow.

The epithelium of the uterus in the intercotyledonary area is not shed during the oestrous cycle and it persists whole throughout pregnancy except in places where the foetal membranes come into close contact with it, when it degenerates and is eroded (see Plate XXIV, 3 and 4, formation of adventitious cotyledons).

The active destruction of the epithelium by the foetal membranes at any time during pregnancy makes it appear unnecessary that shedding of the uterine epithelium should occur during the cycle in order that the foetal membranes may become attached to the uterus. The appearance of the uterus at the end of the 1st month of pregnancy throws light on the changes occurring during the oestrous cycle, as to whether the occurrence of uterine congestion and bleeding just after heat is (1) a preparation for the attachment of the egg shed at that time or whether (2) it is a result of uterine growth initiated by the corpus luteum of the previous ovulation and terminated by failure of the egg shed at this time to develop.

We believe the latter explanation to be the correct one. We have not examined, however, the uterus at the time of the first heat after puberty. Kolster(207) has pointed out that the changes occurring in the uterus during heat are similar to those that are necessary for the development of the foetus, but he did not realise that these conditions disappear again directly after heat and only reappear at a time when the fixation of the foetal membranes is imminent; he examined no very early stages of pregnancy in the cow. Corner(107) found in the sow that in the cycle the time of commencing involution of the corpus luteum (15th–19th day) normally begins a few days after the period at which the foetal membranes become attached to the uterus in pregnant animals (10–15 days). Much the same time relation occurs in heifers; the foetal membranes of one killed at the end of the 1st month of pregnancy had not yet become attached by projections to the cotyledons. In the closely related sheep Assheton (258) from detailed observations found that the blastocyst does not begin to be attached to the uterus until the 17th–18th day after coitus, whereas the next pro-oestrous period would be due about 15 days after; he found that in the very early stages of pregnancy (4th–5th day) the uterus is in the resting stage described by Marshall (54). Thus the changes that occur at oestrus cannot be a preparation for the attachment of the foetal membranes but are rather a result of the previous egg not becoming fertilised and embedded. The menstrual flow and changes in the uterine mucosa of the cow, ewe and sow can have no influence on the fixation of the foetal membranes owing to the time which elapses before fixation occurs. It has been shown that the foetal membranes are capable of removing the uterine epithelium during all stages of pregnancy and there is no necessity to invoke the floating off of pieces of this during the heat period as essential for the attachment which takes place about three weeks afterwards. It may be possible however that the degeneration of the uterine mucosa forms a pabulum for the ovum before attachment. The ample sub-epithelial supply of blood vessels is necessary to supply the needs of the foetal membranes when attachment occurs (see Plates XIII, 4 and XXII, 1), and on failure to attach, due to death or non-fertilisation of the ovum, this supply breaks down and forms the next menstrual flow which occurs in heifers some two days after the next heat.

Table XII shows that heifers which are served by a fertile bull and which become pregnant bleed as usual two days after the heat period in which fertile service occurred; at the time when the next heat period would have occurred if the animal had not been pregnant, however, no blood flow occurred for such uterine flow as may be formed is absorbed by the foetus. This is not all, however, for during the cycle in the cow much vaginal bleeding occurs and the foetal membranes can play no part in absorbing this, so that it can be said that the ultimate cause of the bleeding in the cycle is not merely due to the hypertrophy of the blood vessels in preparation for the fixation of the egg which is caused by the development of the corpus luteum, but is due to atrophy of the corpus luteum which causes degenerative changes in the previously well nourished and vascular mucosa. Oliver (296) has pointed out that in women pregnancy is dated from the first menstrual period missed and that this corresponds approximately with the attachment of the egg in whichever period of the cycle insemination occurs. In extra-uterine pregnancy menstruation occurs at full term although the foetus is dead in the abdominal cavity.

Another explanation of the blood flow in the cow is that the usual pro-oestrous changes have been delayed until after oestrus and ovulation in view of the lateness of attachment in order to nourish the ovum at a very critical stage. If this is postulated it is difficult to relate "the pro-oestrous changes" as due to the ripening of the follicle for in the cow bleeding only occurs after the follicle has ruptured, although admittedly the changes begin before the egg is shed.

The Cotyledons. In the non-pregnant animal these are arranged in longitudinal rows and are small and button like; they increase enormously in size during pregnancy and frequently reach the size of an orange or rather potato, for they become elliptical in shape. Hilty (205) has given measurements to show the size (maximum) of the cotyledons at different stages of pregnancy and figures taken from his paper are shown on p. 53. They indicate that the size of the cotyledon increases regularly throughout pregnancy but that the increase in the latter part consists more of length and height increase than of breadth increase and at this stage the cotyledon is elliptical and not round. The involution of the cotyledon after parturition is described on p. 92. The stimulus for their growth seems to originate mainly in the contact with the foetal membranes, for in heifers killed during the 2nd and 3rd months of pregnancy, when the membranes have not yet become completely attached, the cotyledons in the non-pregnant horn do not appear to develop until pressure is exerted on them by the turgid membranes. Their size is greatest in those parts—near the middle of the foetal sacs—which first come into contact with the membranes in this way and become smaller as the ends of the blastocyst are approached. In the heifer 8 months pregnant the number of the cotyledons attached to the foetal membranes was counted and found to be 63 in the pregnant horn and 41 in the non-pregnant horn; in another cow far advanced in pregnancy there were 53 in the pregnant horn and only 23 in the non-pregnant horn. Bergmann (257) found that not only did the weight of the cotyledons increase during pregnancy but that those of the non-pregnant horn were always smaller than those of the pregnant side, as the following averages from his results show:

Average weight of Cotyledons (gm.).

	Month of pregnancy								
	2	3	4	5	6	7	8	9	10
Pregnant horn	0·2	4·6	17·8	34·4	39·7	109·0	145·4	212·0	255·0
Non-pregnant horn	—	0·7	2·5	5·4	16·7	34·7	57·5	66·0	41·0

At the end of the first month of pregnancy the foetal membranes have not yet become attached to the cotyledons and the blastocyst is held in place by the accumulation of fluid in the central portion (Plate XXV, 1) causing fixation, as Assheton describes in the sheep, in much the same way as the inner tube of a pneumatic tyre is held in the cover. Attachment to the cotyledons by projections from the foetal membranes occurs before the end of the 2nd month (Plate XXV, 2).

In the literature (see pp. 87 and 141) the crypts of the cotyledons are often referred to as though they were formed structures rather than cavities eroded by projections of the foetal membranes round which the connective tissue of the cotyledon has grown. Wall (206), who believes that the crypts are formed structures, remarks with surprise that there is a branch of the chorion for every crypt. There is not room beneath the surface of the non-pregnant cotyledon for the development which occurs in pregnancy and the crypts are formed by upgrowths of the connective tissue round the foetal villi. Craig (17) states that uterine glands open on to the surface of the cotyledons, but this is erroneous. The formation of uterine milk by the crypts of the cotyledons has been referred to above (p. 141). The separation of the foetal from the maternal parts of the cotyledons has been figured by Williams (84) (his figure 78). The foetal part is usually red in colour owing to its large vascular supply while the maternal portion is whitish due to the large amount of connective tissue which it contains.

The histological changes which occur in the cotyledon during the course of pregnancy may now be described. Reference to the description of the virgin condition (p. 87) will show that the cotyledon is covered by a layer of epithelium and is gland-free; no formed elements or crypts appear in it but it consists of an interlacing mass of dense connective tissue and blood vessels. The degree of attachment reached between the 2nd and 4th months of pregnancy will vary materially according to the position of the cotyledon, whether in the centre of the foetal membranes or at the end of the blastocyst; we have however attempted to show how the attachment takes place by giving average stages at each age rather than by describing the various stages existing in one animal.

At the end of the first month of pregnancy the foetal membranes lie over the surface of the cotyledons but no projections are seen interlocking the two (Plate XXII, 1). Where the foetal membranes come into direct contact with the tops of the cotyledons the epithelium is completely denuded, and at the bases of the cotyledons transitional stages from the normal epithelium of the intercotyledonary area through various

PLATE XXII

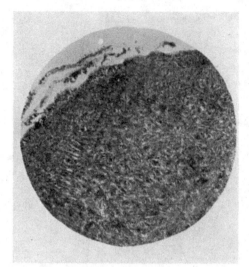

1. Cotyledon: pregnant 1 month.

2. Cotyledon (left side): pregnant 2 months.

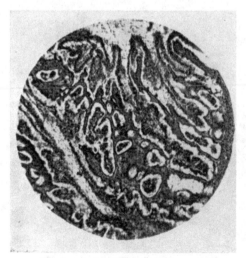

3. Cotyledon (right side): pregnant
2 months.

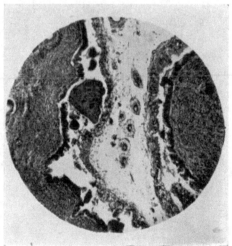

4. Uterine secretions between foetal
membranes and mucosa: pregnant
4 months.

degenerative stages to complete removal can be traced. The stroma of the cotyledon is thickened and dense and contains numerous blood capillaries; extravasated red blood corpuscles can be seen at the tops of the cotyledons next the foetal membranes where no doubt the foetal tissues are beginning to break down and feed upon the maternal.

In the 2nd month the foetal membranes have become fairly firmly attached to the cotyledons by finger-like down-growths which eat their way into the tissues of the cotyledon. The maternal cotyledon responds to this stimulus by growth of the connective tissue and interlocking lamellae are formed (Plate XXII, 2 and 3). It has been shown that before this interlocking begins the epithelium is completely denuded and now no sign of it remains over the cotyledon nor does it reappear in the crypts (see Plate XXII, 2). At the edges of the cotyledons bordering the intercotyledonary or glandular area where the foetal membranes dip down but do not press on the epithelium (which is here intact) large globular cells occur in the foetal tissues just as they do in the crypts and the evidence is all in favour of those cells being of foetal and not maternal origin as Kolster [207] supposes. The larger of the finger-like projections of the foetal membranes contain a central core of mesoblast which is rich in blood vessels and the outer epiblastic layers or trophoblast consists of large cells which here and there have become very swollen and globular. The maternal projections consist of connective tissue only with numerous blood capillaries. At this stage masses of uterine secretion (uterine milk) can be seen between the foetal membranes and the uterine wall in the intercotyledonary area.

At 3 months a larger and more complete system of branching in both the maternal and foetal projections occurs, which increases as pregnancy advances; the relation of the parts and appearance however is essentially the same as that described above. Occasional binuclear cells could be seen in the foetal epiblast.

At the 4th month brownish pigment granules were marked in the foetal membranes just below the epiblast; this pigment is apparently formed from the maternal blood which is set free by the erosion of the maternal tissues and is absorbed by the foetal epiblast.

At the 5th month the collection of maternal red blood corpuscles and other detritus (uterine milk) was very marked between the sacculated parts of the foetal membranes at the top of the cotyledon and the apical ends of the maternal projections. We agree with both Ledermann's [297] and Kolster's [207] description that no lacunae of blood are to be seen surrounding the younger parts of the foetal projections which lie deep down in the maternal tissues, but any maternal blood which is allowed to escape as a result of their erosion appears to collect in the above-mentioned sacculated parts of the foetal membranes on the outer edge of the cotyledon (Plate XXIII, 1); this extravasated blood is there absorbed giving rise to the pigment granules in the course of its destruction. Little if any pigment is to be found in the younger and more deeply situated parts of the foetal projections but patches of these granules were numerous in the foetal epiblast on the outer edges of the cotyledons.

At 6 months practically all stages of the absorption of uterine milk and red blood corpuscles could be observed and Plate XXIII, 2 shows (under high power) masses of blood and detritus collected between the sacculated portion of the foetal membranes on the outer edge of the cotyledon and the ends of the maternal projections. Whole blood corpuscles, broken down corpuscles, amorphous masses of brown pigment and smaller darker pigment granules could be observed in the large cells of the foetal epiblast. By this time the foetal projections have become enormously increased in size and at the outer edge of the cotyledon have a very thick core of foetal mesoblast in which blood vessels and capillaries are large and numerous. The maternal area at the base of the cotyledon between it and the muscular layer of the uterus contains numerous large blood vessels which branch and supply the maternal projections with a plentiful blood supply.

At 7 months the appearance of the tissues is much as before; patches of pigment were also seen in the foetal mesoblast as well as in the epiblast. Blood vessels were increasingly numerous in the foetal projections.

At 8 months the blood supply of the foetal projections was still further increased and at this stage the appearance was essentially similar to that described above, but with even more blood in the sacculated foetal membranes on the outer edge of the cotyledons and more pigment in both foetal epiblast and mesoblast.

It has been shown above (p. 88) that during the cycle pigment is formed in the uterine tissues from extravasated red blood corpuscles; during pregnancy however the extravasated maternal blood is absorbed by the foetal tissues and the pigment is found almost entirely in this position and not in the maternal parts. Kolster(207) has pointed out that in this way iron is obtained for the nutrition of the foetus and Jenkinson(295) has shown that the red blood corpuscles are absorbed by phagocytic action of the trophoblastic cells and are broken down to an iron-containing substance which is transported and a pigment mass

PLATE XXIII

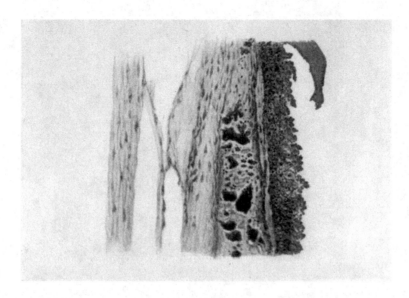

2. Foetal membranes over the cotyledon: 6th month of pregnancy.

1. Cotyledon: 5th month of pregnancy.

which is deposited in the cell; by absorption spectra he found a new haemoglobin derivative—haematophaein—in the foetal placenta.

Kolster (207) has described the large rounded cells in the foetal epiblast (see Plate XXIV, 2) as large wandering leucocytes from the maternal tissue but their frequent binuclear state inclines us to believe that they are cells of the semi-plasmodal foetal epiblast which have become differentiated and filled with a clear mucin-like substance in much the same way as the epithelial cells in the pustules of the amniotic sac (see p. 162). Jenkinson (295), who also figures these cells, does not agree with Kolster as to their origin. Assheton (258) says that in the sheep and the cow they are undoubtedly of foetal origin and in this we agree, for intermediate stages can be observed between them and the normal type of foetal epiblast cell and they are found in the foetal membranes over the intercotyledonary area where it is not applied to the maternal tissues. Jenkinson also describes cells like goblet cells in the trophoblast but was unable to stain them with mucicarmin.

We conclude that the placenta in the cotyledon is formed by the projections of the foetal epiblast eating their way by phagocytic action into the maternal connective tissue of the cotyledon; the latter responds by increased growth and an interlocking system of lamellae is formed. The tip of the foetal projection is the growing part and this lies buried deep in the maternal cotyledon and by its phagocytic action eats into the maternal connective tissue and causes breakdown of maternal capillaries; the blood which is set free together with the highly nutrient lymph (uterine milk) oozes its way to the outer part of the cotyledons where it collects in the sacculated foetal membranes and is absorbed by the foetal epiblast, pigment formation occurring in the process. The difference in the nature of the foetal epiblast in the two different parts of the cotyledon—superficial and deep—is well illustrated by Plate XXIV, 1 and 2 respectively.

Pomayer (298) states that the ends of the foetal projections are situated in maternal lacunae, but the figures which he gives to support this statement show no blood or débris in this space and we conclude that the cavity is caused by shrinkage of the foetal tissues in the preparation of the slide. The accumulation of blood in the spaces between the bases of the foetal villi and apices of the maternal villi is figured by Fraenkel (299) in the cow; he states that the uterine epithelium has either quite disappeared or is in a stage of degeneration. Jenkinson (295) is of the opinion that the maternal crypts both in the adventitious and in the regular cotyledons are formed by growth and budding of the uterine epithelium before projections appear from the foetal membranes. In criticism of his conclusions we can say that we have observed in the oestrous cycle no inclination of the epithelium covering the cotyledon either to bud or form downgrowths which would be likely to happen in preparation for fixation if his theory were correct; on the contrary, the epithelium is frequently cast off by the accumulation of sub-epithelial blood such as Assheton (258) has described in the sheep. Jenkinson does not appear to have examined the cotyledons in the early stages of pregnancy; on his theory the crypt is a downgrowth from the maternal epithelium into the sub-epithelial layer and not as we suppose an upgrowth of the connective tissue of the latter area. In the cotyledon before pregnancy there is not room below the surface for all the development which goes on after pregnancy begins and the tissues of the maternal parts of the cotyledons develop by growing up into the cavity of the uterus and not by downgrowths from it. It is unlikely that the trophoblastic cells, which he himself admits are phagocytic, would fail to absorb the delicate cells of the uterine epithelium and this has been shown to occur by Assheton (258) in the sheep. In this species Assheton brings forward many reasons why the cells lining the crypts are not uterine epithelial cells, although he is doubtful about their origin in the cow; in the latter species however he examined only one stage of pregnancy (about 12 weeks) and our description given above is essentially the same as he describes in the sheep.

Pomayer (298) describes in the foetal epiblast of the cow's cotyledon many double nucleated cells and states that the maternal projections are covered by a layer of flat epithelial cells; we agree with the former but think that the latter, which are also described by Ledermann (297), Jenkinson (295) and Kolster (207), are not true epithelial cells, as they believe, or foetal cells, as Assheton (258) thought in the sheep, but consist of connective tissue plasma or lamellar cells whose function is nutritional and which tend to congregate on the abraded surface. Schafer (300) states that lamellar cells may form an epithelial layer in a tissue or upon its surface. Heape (301) found in a monkey that after denudation of the epithelium by menstruation there was a transference of elements of the stroma to the epithelium, and Schmaus (302) states that in the healing of wounds granulation tissue appears with many fibroblasts or connective tissue cells the largest of which resemble epithelium. In the dog and other similar types there is destruction of the uterine epithelium in the formation of the placenta and we believe that this also occurs in the ungulate placenta; there is no difference between the deciduate and indeciduate type of placenta in this respect such as is described by Jenkinson (303).

Sommer (115) describes a yellowish fat infiltration of the uterine mucosa about 3 days after calving and Wall (206) a waxy degeneration 5–14 days post-partum. That the separation of the foetal membranes at birth is not caused by fatty degeneration of the cells of the cotyledons is shown by the fact that Ledermann (297) found fat granules both in the foetal epiblast and in the cells which he describes as maternal epithelium in the early stages of pregnancy. Kolster (207) has also described fat globules in the so-called epithelial cells of the maternal cotyledon during the early stages of pregnancy in the cow. It has been shown above (see p. 76) that the connective tissue lamellar cells play a part in the transference of lipoids to the Graafian follicle and corpus luteum and the accumulation of fat in similar cells in the cotyledon after parturition, as found by Hilty (205), is the natural result of its non-utilisation by the foetus. Williams (286) found that there was no loosening of the membranes before birth but on the contrary the firmness of adhesion of the membranes to the uterus increases with the period of pregnancy. We have not investigated the causes which lead to the freeing of the foetal membranes after birth, but from a study of the structure of the cotyledon and the way in which separation can be made during pregnancy conclude with Pomayer (298) that contraction of the muscular coats of the uterus is an essential factor. Pressure at the base of the cotyledon will have the same effect in forcing out the foetal parts as traction on the foetal membranes themselves; the freeing is no doubt greatly assisted by the uterine milk or lochia from the maternal part which floats off the foetal tissues, and also the shrivelling of the foetal portion caused by the loss of blood from its tissues as a result of the birth of the foetus is a further aid (Assheton (258)). Sommer (115) found that in the cow the afterbirth comes away normally 1½–8 hours post partum, on the average 4½ hours, and Romolotti (304) found it to be 5 hours. As Pomayer (298) has pointed out retention of the afterbirth is most frequent in cases of uterine inertia due to infection, debility or premature birth. Williams (286) states that the tendency for retained afterbirth increases with the duration of pregnancy before abortion so that lack of actual muscular development cannot be the cause in this case. Pressure caused by contraction of the uterus at parturition is without effect in detaching the cotyledons owing to the presence of the foetus against which the pressure is made; it is also ineffective while the foetal villi are distended with foetal blood. Williams (286) states that when abortion occurs before the foetus is 30 cm. long it is expelled enclosed in the membranes. Pomayer (298) concluded from histological examination that retention of the foetal membranes may occur as the result of congestion of the blood vessels which causes close interlocking of the tissues. Muscular contraction of the uterus would naturally reduce congestion and the administration of purges, which has the same effect, is known to assist the shedding of the afterbirth. That the degeneration of the foetal tissues plays a part is seen by the greater ease with which the membranes are separated when the uterus has been kept some time after the animal has been killed. Pomayer states that under pathological conditions the epithelial tissues may degenerate and the foetal and maternal connective tissue grow together; this is likely to occur in cases of infection with *Bacillus abortus*, although Lehmann (305) attributes retention under these conditions mainly to the oedema and swelling of the tissues that occurs as the result of infection. Wall (206), on the other hand, states that infection by *Bacillus abortus* seems to loosen rather than tighten the connections in the cotyledons and he attributes retention after abortion to secondary infections which cause paralysis of the muscular coats and inflammation in the cotyledons.

The period of the oestrous cycle at which attachment of the foetal membranes occurs has been described above (p. 145). Since it has been found by Fraenkel (42) in rabbits that the fertilised egg does not become attached if the corpus luteum is destroyed and also that destruction of the corpus luteum after attachment results in the death of the foetus, it must be concluded that the corpus luteum produces an internal secretion which sensitises the uterine mucosa and makes it respond to the pressure of the foetal membranes. Loeb (47) has shown in the guinea-pig and Hammond (208) and Nielsen (90) in the rabbit that stimulation of the uterine mucosa by wounding it during a period when the internal secretions of the corpus luteum are active causes the formation of the decidual cells of the maternal placenta. Krainz (260), however, in the bitch found structural changes as the result of this operation but no formation of deciduomata; in this animal however the placenta is normally produced without the formation of large decidua and the responses to the wound stimulus in different species will naturally be those that normally occur on fixation of the membranes.

That the placenta itself is formed by the continuous active erosion of the foetal membranes is shown by the fact that the experimental deciduomata do not go on increasing in size after the first growth nor does an inert foreign body give rise to more than one crop of deciduomata (see references p. 126). Moreover the unfertilised egg, itself a foreign body but with no power of growth or erosion, does not stimulate a reaction by the uterine mucosa. Transplantation experiments with ovaries and other tissues have shown that the graft is much more likely to take when tissues from a young animal are transplanted into an old one than *vice versa*, by reason of the greater growth power of young tissues which tend to erode and live upon the older tissues; the same principle applies to foetal tissues in the uterine mucosa of the mother.

Fellner (306), Schröder and Goerbig (307) claim to have stimulated growth of the uterus by injection of placenta into young virgin rabbits, but in our experience the size of the uterus in such animals is normally very variable owing to variations in the rate of maturity, and we do not consider these experiments conclusive.

Adventitious Cotyledons. Sometimes before, but most frequently at and after, the 4th month of pregnancy the foetal membranes often become attached to the uterus in places other than those of the regular cotyledons. These adventitious cotyledons are formed in certain places where the foetal membranes come into close contact with the uterine wall; the foetal membranes in these places are only lightly attached to the uterine wall and the membranes can be pulled apart much easier than in the true cotyledons. The number of these adventitious cotyledons generally increases as pregnancy advances; they occur mainly in the body of the uterus. Craig (17) says they are more frequent near the os and Franck-Albrecht (244) figures them as frequent round the entrance to the cervix; in this area the pressure of the foetal membranes on the uterus is usually greatest, as will be seen from the contractions appearing in the foetal membranes (Plate XXV, 3).

From a comparison of the number of cotyledonary attachments in the pregnant and non-pregnant horns of the uterus it is apparent that the power, not only of developing the dormant cotyledons of the uterus, but also of initiating the formation of new adventitious cotyledonary growths rests with the foetal membranes.

Rörik (273) concluded that the formation of accessory placentae in the cow is due to the insufficiency of the normal cotyledonary attachments, for he found that in early pregnancy 1 gm. of foetus had a cotyledonary surface of 13–17 sq. cm. whereas in late pregnancy the surface was reduced to 7–11 sq. cm. This however cannot be the cause, for the non-pregnant horn of the uterus contains many true cotyledons which do not become functional during pregnancy, as may be seen by the difference in number between the cotyledonary attachments of the pregnant and non-pregnant horns (see p. 147).

In the related horse (see Ewart (263) and Kolster (293)) and pig (see Assheton (258)) the foetal membranes produce projections and become attached all over their surface and not merely in certain areas as in the cow and sheep. Wall (206) has described villi 100–200 μ long—on the foetal membranes of the cow in the intercotyledonary area (at the 7th–9th month) of pregnancy. Since at this stage the uterine mucosa is wrinkled it is probable that these projections fit into the grooves so formed. In the cow the presence of raised projections—the cotyledons—causes greater pressure on the foetal membranes in certain areas more than in the other parts of the uterus; moreover the cotyledonary area is gland-free and so there is no secretion here to prevent close apposition of the parts such as exists in the intercotyledonary area where the secretion poured out from the mouths of the glands tends to float off the foetal membranes as well as to supply this part of the foetal tissues with nourishment. Sometimes, however, owing to the scarcity of glands and the pressure of the foetal fluids or of the foetus itself, the foetal membranes become closely pressed to the intercotyledonary area. When this happens the uterine epithelium is eroded and the dense sub-epithelial layer of the uterine mucosa thickens; such an attachment in the inter-cotyledonary area is shown in Plate XXIV, 3 and forms the beginning of an adventitious cotyledon. Hilty (205) found that adventitious cotyledons were formed from this sub-epithelial layer in the cow in much the same way as in an ordinary cotyledon. The foetal membranes grow projections which eat into the uterine connective tissue and this in turn grows out and an interlocking area is produced (see Plate XXIV, 4); extravasation of blood occurs between maternal and foetal parts and pigment is produced in the foetal part just as in a normal cotyledon. The interlocking and growth however are not so great as in a normal cotyledon and the attachment is easily pulled apart (but not so easily as it is in the pig) for the projections are straight and not many secondary branches are formed. The proliferation of the sub-epithelial mucosa in the intercotyledonary area in the cow is not dissimilar to that figured by Kolster (293) in the mare where the placentation is diffuse.

(3) *Foetal Tissues.*

The appearance of the foetal epidermal appendages has frequently been used as a method of determining the age. Bergmann (257) found in the "rotebunte Niederungs" breed that the first hairs appear round the

PLATE XXIV

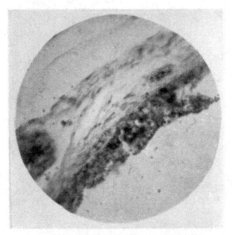

1. Foetal membrane on outer edge of
cotyledon: 8th month of pregnancy.

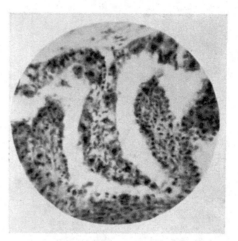

2. Foetal membrane on inner (uterine)
side of cotyledon: 8th month of pregnancy.

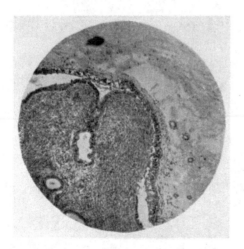

3. Erosion of intercotyledonary epithelium by
foetal membrane prior to formation of adventitious cotyledon: 2nd month of pregnancy.

4. Adventitious cotyledon: 6th
month of pregnancy.

Table XXIX. *Some changes occurring in the foetus and its membranes during pregnancy.*

Month (4 week) of pregnancy	No. of heifer	*Foetal length in cm.	Foetus Organ development	Foetus Hair development	Foetal membranes End of membranes	Foetal membranes Cotyledonary attachments	Foetal membranes Amniotic pustules
1	P 2	0·8	Limb sprouts first forming	Nil	Centre of membrane distended with fluid (18 cm. long). Ends long (each 18 cm.) and narrow without fluid contents	Nil	Nil
2	P 4	5·4	Ear bud just forming. Eye dark spot just under surface. Digits quite distinct. Mouth and nostril depressions formed	Nil	Moderately long projection of non-fluid-containing and non-vascular part extending beyond the doubled over fluid-containing and vascular chorionic part	Well marked projections into cotyledons within about 5 inches of navel. Beyond this shadowy outlines over cotyledons and at ends none at all	Nil
3	P 3	13·3	Periople ring and top of hoof swollen. Hornbuds visible. Ridges on inside of ears	Roughness round muzzle and lower lip	As before, non-fluid-containing end hard and tough	Well marked in pregnant horn of uterus. Shadowy in non-pregnant horn. Constriction in body of uterus and outlines of cervical mucous plug seen	Small ones present
4	P 1	24·5	Eye sunk deeper under surface; pigment not so distinct as before	Roughness round muzzle and lower lip	As before	Projections of cotyledons in both horns of uterus but not so large on non-pregnant side. Dark pigment round cotyledons and round adventitious attachments	Many both large and small
5	P 5	39	As before	Long hairs round muzzle and lower lip and on eyebrows	As before	Cotyledons in non-pregnant horn smaller than in pregnant side but both large	Very large and numerous; coalesce in places
6	P 7	51	As before	As before, also downy hairs over horn cores and roughness at end of tail	Hard shrivelled lump at end of membranes	As before, larger	Smaller and shrivelled
7	P 6	68	Eye open. Meconium from anus	As before and also hair on tail, eyelashes and round eyes and inside ears; long over horn cores and round hoofs and in decreasing amounts to top of leg. Hair round navel and on elbow joint	As before	As before, larger. Large number of adventitious cotyledons	Very shrivelled; have almost disappeared
8	P 9	80	Hard translucent horn on hoofs. Meconium from anus	Hair all over body. Colour (roan) quite distinct	Thin projection about 3 inches long without fluid	As before, larger. Whole uterine surface brownish	Practically disappeared; few shrivelled threads left

* Forehead to base of tail.

muzzle and eyebrows at the end of the 4th month of pregnancy, while at the 6th month hairs also appear on the eyelids, ears and over the horn cores; at the 7th month they are visible on the tail and at the beginning of the 8th all the body is covered with short hair.

Our series of crossbred Shorthorns (see Table XXIX) show that at the end of the 1st (4 weeks) month the limb sprouts are just forming; at 2 months the ear bud can be seen and the digits are quite distinct. At 3 months the horn buds and ridges inside the ears are visible while the skin round the muzzle and lower lip is slightly rough but no hairs have yet appeared. At 4 months the foetus has much the same appearance but the pigment of the eye which has hitherto been prominent is now more obscured by this organ sinking deeper under the surface. At 5 months long hairs appear round the muzzle, lower lip and on the eyebrows. In addition at the 6th month downy hairs appear over the horn cores and the skin appears rough at the end of the tail. At the 7th month hair makes its appearance on the tail, eyelashes and round the eyes which are now open; hair also appears on the inside of the ears and is long round the periople ring of the hoof and is present but in decreasing amounts up to the top of the leg. Meconium appears from the anus. At the 8th month hair is present all over the body and the colour (roan) is distinct; hard translucent horn appears on the hoofs and meconium comes from the anus. The extent to which the translucent horn develops on the hoof has also been used as a means of determining the age of a calf after birth by Schultz[308] and Schwarz[309] for meat inspection purposes. How far these appearances are affected by breed and nutritional conditions is not known but they will no doubt be of use in estimating the approximate age of any foetus obtained. Müller[310] has described the appearance of the long hairs of the muzzle, etc. (Sinus hairs) in calf foetuses of known lengths, but, in most cases, of unknown ages; he found none at the 2nd month of pregnancy but they appeared soon after this time.

At the end of the 1st month of pregnancy although the foetus itself is yet small (about the size of a pea) the foetal membranes are quite long (54 cm.) and extend throughout both horns of the uterus (see Table XXIX). The foetus itself lies on one side—the pregnant horn— of the uterus and in this the more central parts of the foetal membranes have become filled with fluid (18 cm. long); the apical parts of the foetal membranes have little fluid in them and consist of long rounded strands (Plate XXV, 1) like the early stages in the sheep described by Assheton; no attachments to the cotyledons are visible. Albrechtsen[120] states that

PLATE XXV

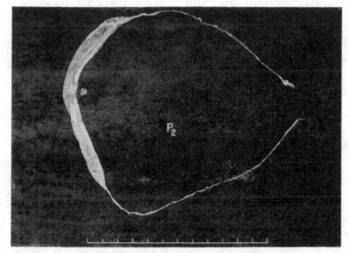

1. Foetus and foetal membranes at 1 month.

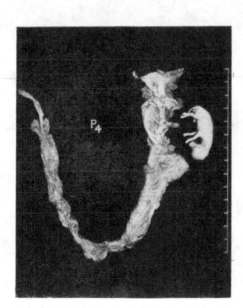

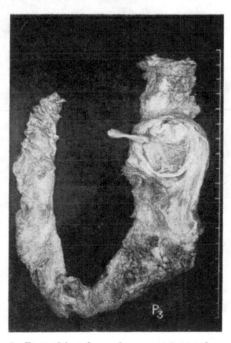

2. Foetus and part of foetal membranes
at 2 months.

3. Part of foetal membranes at 3 months.

during the 1st month the foetus is loose in the uterine horns and is nourished only by diffusion from the uterine milk. Schmaltz[1] has figured an early stage (1 cm. long) in the cow showing the foetal membranes undistended with fluid; he also states that at about 7 weeks (3·5 cm. long) the foetal membranes are not yet attached to the uterus. At the end of the 2nd month the extreme ends of the foetal membranes are still undistended with fluid but these are now comparatively short and the distended chorionic parts are seen to be folded and growing over the non-vascular end. Cotyledonary villi are only seen in the more central parts of the membranes and shadowy outlines of the cotyledons can be observed over the area next this part; towards the end of the distended portion of the membranes the surface is quite smooth (Plate XXV, 2). Williams[286] states that the foetal membranes are not attached to the uterus until the embryo is 5–8 cm. long; which, according to our measurements, is between the 2nd and 3rd months. At the end of the 3rd month of pregnancy the ends of the foetal membranes still have a short undistended narrow portion which persists up to at any rate the 8th month; this end has now become hard and tough. The cotyledonary projections are quite marked in the pregnant horn but in the non-pregnant horn shady outlines of the cotyledons only are to be seen. Where the membranes pass through the body of the uterus they are slightly constricted and the round outline of the mucous plug of the cervix can be seen impressed on the surface of the membranes (Plate XXV, 3). Williams[84] (his Plate II, p. 124) has figured the foetal membranes at about this age and states that the extreme ends often show a necrosis which is usually greater in the non-pregnant horn; he states[286] that in the non-pregnant horn the end usually extends to the Fallopian tube as a hard flattened yellow cord but on the pregnant side it usually becomes invaginated into the cavity of the amnion or allantois. Jenkinson[295] has shown that this necrosis is caused by lack of blood vessels in the extreme parts with subsequent degeneration of the epiblastic layers; a nodular growth of connective tissue with the development of a muscle sphincter serves to prevent escape of the foetal fluids. This terminal atrophy of the membranes is probably a provision for preventing the foetuses in cases of multiple pregnancy from attempting to absorb each other by a growing together of their membranes. The necrosis of the ends of the membranes is not so marked in the cow as it is in the pig or the sheep and this may possibly account for the frequency of freemartins in cattle as compared with sheep or pigs. We have not observed any cases of twin pregnancies in the cow but Lillie[311] states that in such cases he has constantly observed a continuous chorion due to the fusing of the two membranes in their vascular layer; on this fact he bases his theory of the origin of freemartins. Statistics quoted by Lüer(312) and Keller(313) show that 95 per cent. of heifers twin to bull calves are sterile.

At the end of the 4th month of pregnancy the cotyledonary attachments are seen in both pregnant and non-pregnant horns of the uterus although they are not so large in the latter as in the former; at this stage fixation of the membranes may be said to be complete. Dark patches of pigment can now be seen in the foetal cotyledons and round the attachments of the adventitious cotyledons where they have been pulled away from the uterus.

From this time up to the 8th month of pregnancy a similar state of things is to be seen. Bergmann(257) by counting the attachments of the cotyledons concluded that further fixation to the cotyledons ceased after the end of the 3rd month; he found the average number of attachments was 58 in the pregnant horn and 30 in the non-pregnant horn.

It has been shown above (p. 48) that the right ovary is generally more active than the left in the cow and Stålfors(314) found that pregnancy occurs more frequently in the right horn of the uterus than in the left. The foetus usually develops in the horn of the uterus of the side from which the ovum was shed, but Bergmann(257) found cases in which the ovum had migrated to the other side. While migration of the ovum is common in animals having many young at birth it is not frequent in those that bear one, such as the cow. In the cow even in cases of twins no records have been made of migration occurring; Bergmann(257) in two cases of twin pregnancy and Küpfer(89) in 17 cases of twin pregnancy failed to find cases of migration. Whether the migration that occurs in the cow is external or internal is unknown; external migration has been shown to occur in the rabbit by Leopold(315). Since migration is uncommon in cases of single ovulations (cow—Bergmann(257) found 1 in 50 cases and Küpfer(89) 1 in about 1000 cases) but is common in cases of multiple ovulations (sow—Corner(99) 50 per cent., Küpfer(316) 25 per cent.) it would appear that migration may occur at a time when the foetal membranes begin to swell and the young are orientated in the uterus by "touching out" of their ends. Since attachment does not occur for some little time the blastocysts, when the accumulation of foetal fluids begin, would naturally tend to distribute themselves in a single line in the uterus owing to pressure conditions and the undistended ends of the foetal membranes will frequently overlap as they do in the pig. In after-development the parts of the uterus to which the foetus is first attached tend to grow more than those in which no attachment is made; this can be seen by the very unequal lengths of the two horns of the uterus in the pregnant cow. Franck-Albrecht (244) state that in a 9-months pregnant cow's uterus the fertile horn measured 71 cm. whereas the non-fertile one was only 37 cm. long. This lengthening of the uterus is very marked in cows in which both foetuses lie in the same horn. The increase in length of the uterus in itself tends to equalise the distances between the embryos.

The appearance of the foetal fluids up to about the 6th month of pregnancy is usually clear and transparent but is occasionally slightly whitish and turbid. At about this time, however, the allantoic fluids become yellowish, although the amniotic fluid still remains clear. After about the 6th month hippomanes or lumps of a yellowish soapy consistency may be found in the allantoic fluid. These have been shown by Bonnet and Schauder (317) in the mare and Jenkinson (295) in the cow and sheep to arise by the more solid parts of the uterine secretions accumulating in sacculated portions of the foetal membranes from which they are eventually cut off and float free in the fluids. In the sheep they are common and one is shown in Plate XXXIII, 4. From about the 6th month of pregnancy onwards small golden patches like grains of sand can be seen by the naked eye between the foetal membranes and the uterine wall; they could not be dissolved by ether and did not consist of fat but are probably of an amino-acid or protein nature. In sections or smears taken from between the uterine surface and foetal membranes these granules appear as concretions in which large dark-staining masses appear as well as numerous semi-decomposing nuclei scattered in the amorphous mass (Plate XXII, 4). Crew and Fell (318) have described somewhat similar bodies in testes where the products of secretion are undergoing absorption. There can be little doubt that these concretions are formed from the less soluble parts of the uterine secretions and that large accumulations of them give rise to hippomanes (one of which was found in the 8 months pregnant stage) and are the origin in part of the yellow colour of the allantoic fluid. Jenkinson (295) found that the pigment of these bodies was bilirubin, probably derived from decomposition of blood pigment; he also describes the frequent occurrence of calcium oxalate crystals in these bodies as well as in the foetal trophoblast.

Bergmann (257), who made a thorough investigation of the foetal fluids in the cow at different stages of pregnancy, found that the allantoic fluid which is at first clear and colourless becomes yellowish and cloudy and in the last third of pregnancy a yellowish brown to brown colour;

H

11

Reproduction in Cattle

it is always fluid in consistency and the specific gravity varies from 1008–9 at the beginning of pregnancy to 1020–22 at the end; the reaction was alkaline in the first two-thirds of pregnancy but varied in the last third. Nauta (319) has pointed out that in the calf the urachus is passable during all foetal life and that foetal urine mixes with the allantoic fluid. The amniotic fluid Bergmann (257) found was at first clear and colourless and then yellow and clear but often cloudy and in the last third of pregnancy whitish and slimy; the reaction was always alkaline and the specific gravity remained the same—1078–98—throughout.

We found at the 8th month of pregnancy that the amniotic fluid was light yellow in colour and thick with a mucous secretion. This mucus is not present in any great quantity in the early stage of pregnancy and its origin must be sought in the walls of the amnion and epithelium of the foetus. It was noticed that the amniotic portion (in the cow there is no allantoic portion) of the navel cord and internal surface of the amnion was covered in the early stages of pregnancy (3rd month—see Fig. 69) by a number of small whitish pustules. These pustules which become very numerous and turgid during the 4th month of pregnancy (see Table XXIX) reach their maximum size and distribution at the end of the 5th month; after this time they become smaller and more shrivelled and by the 7th–8th month have almost disappeared, only a few still showing as shrivelled threads or lines on the surface of the amnion. The time of their shrivelling and disappearance coincides with the appearance of large quantities of mucus in the amniotic fluid.

Bernard (320) found that the "plaques amniotics" are localised masses of cells on the internal surface of the amnion and contain large quantities of glycogen; he says they reach their full development in the calf at about the 5th month and then gradually atrophy.

We have sectioned the pustules in the different stages of development. At the 3rd month (Plate XXXII, 3) the pustules are seen as a thickened mass of epiblastic cells and the sub-epiblastic layer of rather dense mesoblast in this region is also thickened. The cells of the epiblastic layer are very similar to those of the foetal epidermis, the basal layer is of cubical deeply-staining cells, the middle layers consist of large clear globular cells with a deeply-staining nucleus and the outer layer is of slightly smaller cells rather more deeply staining. At 4 months the pustules have become thicker and the globular cells more numerous; several of the outer layers of cells have now become flattened and shrivelled just as they do in the epidermis of the foetal skin. At the 5th month (Plate XXXII, 4) the size of the whole pustule has enlarged and the proportion of the outer shrivelled layer of cells has increased; the thickened layer of mesoblast lying immediately under the pustule is well supplied with blood vessels. At the 6th month the pustule has become smaller and the proportion of outer shrivelled cells has still further increased. At the 7th month the pustule was small and although many clear cells were present they were smaller, more especially the deeper ones which were more darkly staining than before; blood vessels were very numerous in the thickened mesoblastic layer. At the 8th month the pustule, which was long and thread-like, had an appearance very similar indeed to that of the foetal skin epidermis; distinct layers were present as in the skin, although here and there a large globular cell could be seen; the base adjoining the mesoblast was thrown into papillae and in the mesoblastic portion of these blood vessels were numerous (Plate XXXII, 5).

Jenkinson(295) who has figured the structure of these cells at the 4th-month stage in the cow describes them as vacuolated with glycogen and says that the cell structure may break down and become converted into a bag containing but one large mass of glycogen, and that in the older stages, when the glycogen becomes used up, the cells become flattened.

We have stained these cells with mucicarmin and find that the large globular cells do not take the stain but that the shrivelled cells on the outer side stain slightly. From their origin in the early stages and appearance in the old stages it is evident that the pustules are epidermal thickenings which apparently play no important function in metabolism; their structure is essentially similar to that of the skin of the foetus itself (see in Plate XVII in development of mammary gland). Their store of glycogen differs in no respect from that stored in the foetal tissues. Mucin, although not demonstrable in these cells, is produced in the amniotic fluid and over the surface of the foetal skin; whether it is formed from the stored glycogen is not definitely known although it seems probable as the outer cells collapse at the time of its production and in the stratified epithelial cells of the cow's vagina mucus can be stained along the borders of similar clear epithelial cells (Plate XI, 6).

The production of mucus by the epidermal cells covers the skin of the foetus with slime which is of great assistance in lubricating its passage at birth.

Williams(84) states that in the mare, where the allantois forms a completely enveloping sac, the foetus may be born more or less completely within the amniotic sac; in the cow however an extensive area of the amnion is not covered by the allantoic sac (see Franck-Albrecht's (244) Fig. 76) and causes the former to remain adherent to the chorion and the foetus to be born naked. Moreover in the calf the navel cord is short and usually breaks at birth, whereas that of the mare is longer and does not rupture. In the cow therefore mucus is essential for lubrication of the passage of the foetus at birth and an additional supply is available from the mucous secretion of the cervical plug (see below).

It is usually stated that before birth occurs the foetus is rotated in the uterus but we have made no direct observations on this point.

The foetus which has hitherto been lying on its back owing to the position of its centre of gravity naturally tends to accommodate itself to the outlines of the uterus when uterine contraction begins (Dennhardt (321)). The form of the uterus is convex above and the back of the foetus which is convex naturally fits into it as a result of pressure by the contracting uterus and rotation is produced. If this does not take place difficulties in parturition occur, for the roof of the pelvic passage in the cow is concave and that of the pelvic floor convex.

Whether the presentation is posterior or anterior would seem to depend on the centre of gravity of the foetus, the heavier portion in the cow tending to drop furthest into the body cavity. Since the heavier end of the foetus in the early stages of development is the fore end, while in the latter stages the hind end develops more, it is not unlikely that craniocaudal rotation occurs during development; the spiral arrangement of the blood vessels in the umbilical cord admits of this possibility. Williams (84) states that in the vast majority of cases in the larger domestic animals (mare and cow) anterior presentation occurs at the time of birth, while in the smaller multiparous animals (sow and bitch), where the uterus is not pendulous but lies along the abdominal cavity, the foetus may present in either way but most frequently anteriorly. From this it would appear that the pendulent condition of the uterus is the main cause of the anterior presentation in the cow.

(c) *The Cervix.*

During pregnancy the cervix becomes completely occluded by a plug of mucus of thick sticky consistency, the amount of which increases as pregnancy proceeds. Reference has already been made to the change in consistency from the non-pregnant to pregnant condition (p. 55). Williams (84) states that sometimes the vaginal end of the seal is inconspicuous while within the canal it may be well developed.

PLATE XXVI

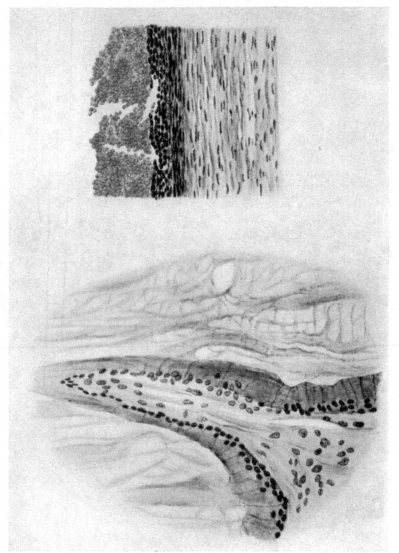

1. Cervical epithelium: 8th month of pregnancy.

2. Penetration of blood through wall of follicle.

The mucus accumulates not only between the lamellae in the cervix (see Plate XIV, 4 and compare with 3, during the cycle) but it also extends to and blocks the central lumen and generally forms a seal over both the internal and external os. It thus effectively shuts off the uterine contents from external infection by .way of the vagina. The accumulation in the lumen and between the lamellae stains intensely with mucicarmin.

Measurement has been made of the thickness of the mucin at the bases of the lamellae in different stages of pregnancy and the following results were obtained (diameter in micrometer squares):

	Month of pregnancy							
	1	2	3	4	5	6	7	8
Diameter	2–3	6–7	11–12	6–7	7–13	18–26	9–20	20–51

The plug is formed by the epithelial cells of the cervix which during pregnancy can be observed in all stages of the secretory process—from high columnar cells filled with mucin to flattened and irregular cells with the mucin straining from them and flowing between the lamellae (Plate XXVI, 1).

It has been shown above (p. 93) that in the oestrous cycle the secretion of thick mucus is apparently associated with the presence of a well-developed corpus luteum, and that liquefaction of this mucus occurs with atrophy of the body. During pregnancy, when the corpus luteum remains large, the accumulation of thick mucus is greatly increased, and just before parturition (see Appendix I (12)) liquefaction of the mucous plug occurs just as it does at a heat period. Zieger(110) states that the flow usually begins about a week before parturition, but that mucus may appear from the vulva at any time after the 5th month of pregnancy. The accumulation of mucus lubricates the vagina for the passage of the foetus at parturition. Williams(286) states that when the foetus has died or is decomposing in the uterus the cervical seal is dissolved; the living foetus acts on the uterine seal, we believe, through its action in maintaining the corpus luteum rather than directly.

Whether the mucous plug in the cervix is developed to a similar degree in other species is not known; but in our experience it is not nearly so well marked in rabbits and pigs as it is in the cow. Trautmann(212) has figured a mucous plug in the cervix of the pregnant ewe in which species the seal is well developed.

Zieger(110) found the reaction of the cervical seal was strongly alkaline and that the mucus was soluble in alkaline salts but became hard in 60 per cent. alcohol.

The cervical seal of mucus besides forming a barrier to bacterial infection of the uterine cavity during pregnancy (which accounts for the difficulty in causing abortion by injections of *Bacillus abortus* per vaginam during pregnancy) also probably helps to dilate the cervix in readiness for the approaching parturition; and its liquefaction before the onset allows the foetal membranes to penetrate the cervix and still further dilate it. Zieger(110) found that at the 7th month of pregnancy in the cow the external os would allow the passage of a finger, which is quite impossible during the cycle or early stages of pregnancy. Servatius(117) states that a day after parturition the cervical canal is closed by a thick slime and remains so until involution of the uterus is complete. Sommer(115), who studied the involution of the cervix after calving, found that it usually became closed and the folds in the mucosa reappeared 8–14 days post partum, but this took longer in old than in young animals; he found that the internal os closes first and the external os later, and that the latter at 3–4 weeks after calving will usually admit one finger.

(d) *The Vagina.*

Oppermann (239) states that during pregnancy the vaginal walls become dryer and more sticky to the touch. During pregnancy the upper end of the vagina next the os undergoes very similar changes to those occurring in the cervix, and the folds of the epithelium are coated with a thick layer of sticky mucus. In sections stained with mucicarmin free mucus can be seen in the bases of the folds as well as in the epithelial cells (see Plate XXVII, 4).

Retterer and Lelièvre (322) found and have figured the development of mucus in the outer layers of cells of the stratified vaginal epithelium of guinea-pigs during pregnancy.

The appearance of the vagina above the urethra is very similar to that which occurs about three days before the onset of heat during the cycle (see p. 95); there is no sub-epithelial congestion, but lymphatic nodules are occasionally seen.

In the part of the vagina next the vulva the appearance during pregnancy is again very similar to that which occurs in the cycle about three days before heat is due (see Plate XV, 1 and p. 96). In addiiton, however, the lymphatic nodules appear to be particularly well developed, more especially in the earlier months of pregnancy. One of these lymphatic nodules is shown in Plate XXVII, 5. In granular vaginitis these lymphatic nodules enlarge considerably.

PLATE XXVII

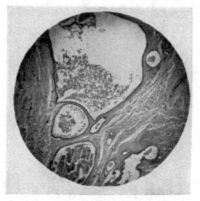

1. Gärtner's canal from cow: cystic.

2. Gärtner's canal: 3rd month of pregnancy.

3. Gärtner's canal: 48 hours after beginning of heat.

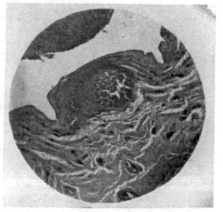

4. Vagina next os: 8th month of pregnancy.

5. Lymphatic nodule in vaginal mucosa: 1st month of pregnancy.

Just before parturition, at a time when the flow of mucus begins, the vulva and vagina become red and swollen (see Appendix I (12)), but the histological changes occurring at this time have not been investigated.

Vaginal secretions. The effect of pregnancy on the flow of blood from the vulva after the heat period in heifers has been closely observed, for it is a common belief among herdsmen that bleeding does not take place two days after heat if the heifers have been successfully served by the bull and become pregnant. Williams[236] states that if a cow has been served and conceived it is doubtful if there will be a menstruation following; it is an early and valuable sign that conception has occurred.

Four heifers were selected which bled regularly and freely 2–3 days after their heat periods (P 5, P 6, P 7 and P 8, see Table XII); these were then mated with a fertile bull, two early and two late in the heat period. All except one (P 8 which was served late and afterwards several times failed to conceive) became pregnant to this service. In all bleeding took place after fertile service at the normal time (2–3 days), but subsequent bleeding was prevented in the three which became pregnant, but not in the one which was sterile.

These facts show that bleeding is essentially part of the last oestrous cycle. One can say that the pro-oestrous bleeding in the cow is telescoped so as to appear after the oestrus, but if so it is difficult to explain the ovarian changes as the direct cause of the uterine changes, for different time relations between the two occur in the bitch and the cow.

The results are interesting in connection with menstruation in women and in the bitch. In the cow insemination before the menstrual flow results in fertilisation of the ovum; in the bitch insemination occurs after the flow. As to whether in women insemination occurring just before menstruation is successful, as in the cow, there is not much evidence; but the majority of writers believe that this is so, for at no definite stage in the cycle is insemination known to be unfertile (but see Siegel[68]). Woman, however, differs from the cow in the time relations between ovulation and menstruation (see p. 27).

That pregnancy can occur before the first menstrual flow in women, and that conception frequently occurs during the amenorrhœa of lactation (Vignes[188]) points to the fact that the menstrual flow is the termination rather than the beginning of the oestrous cycle.

That bleeding does not mark the initial stages of a new cycle, but rather the termination of the old cycle is difficult of experimental proof, for in the cow the cycles overlap. An attempt which was made to separate the two cycles, by pricking the ripening follicle per rectum, and thus preventing the immediate onset of the new cycle, failed owing to the attempt being made on an old cow, the ovaries of which were too overgrown with connective tissue to locate properly the ripening follicle per rectum, and which we failed to rupture by the operation. No opportunity was obtained for repeating the experiment in young virgin heifers which bled regularly after their heat periods.

During pregnancy there are no changes in the character of the cellular contents of the vaginal smears which are distinctive. Normal and degenerate vaginal epithelial cells are present, and occasionally a polymorphonuclear leucocyte or two; in the upper regions of the vagina sticky mucus is abundant.

Gärtner's canals. The structure of these has been described on p. 97. Sections obtained at the 3rd, 4th and 6th months of pregnancy showed a far greater development of the gland tubules and accumulation of the secretory products in the lumen than has been observed in sections obtained from heifers during the oestrous cycle (compare Plate XXVII, 2 pregnant with 3, during the cycle).

Roeder(221) has shown that the size of the canals varies greatly in different individuals. It may be that the differences we have observed were due to the small number of individuals examined (4 or 5 of each state), or to differences in the part of the canal sectioned, whether apical or near the exit; but we are inclined to believe that this was not the case, and that hypertrophy occurs during pregnancy.

(e) *The Mammary Glands.*

The changes that occur in the gross weight of the udder in heifers (all of about the same age and type) at different stages during their first pregnancy, are shown in Table XXX.

It has already been shown (p. 66) that the udder in virgin heifers collected from slaughterhouses varied in weight from 1420 gm. at 2 years to 2650 gm. at 3 years, averaging 1930 gm. at 2 years 3 months old. The average weight of the udder of 7 heifers which were killed during the oestrous cycle (and which were similar in type but slightly younger than those which became pregnant) was 2250 gm. and varied from 1240 to 2740 gm.

From Table XXX it will be seen that very little increase in weight of the udder takes place in the early stages of pregnancy, but that it gains in weight more and more each month as pregnancy proceeds. The number of animals on which this statement is based is admittedly

PLATE XXVIII

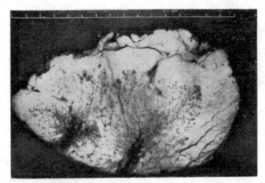

1. Udder of heifer: 3 months pregnant.

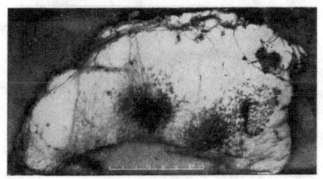

2. Udder of heifer: 4 months pregnant.

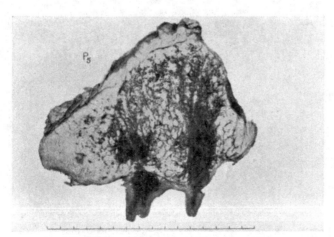

3. Udder of heifer: 5 months pregnant.

Table XXX. *Weight changes in the udders of heifers during the first pregnancy.*

Month (4 week) of pregnancy	1	2	3	4	5	6	7	8
No. of heifer	P 2	P 4	P 3	P 1	P 5	P 7	P 6	P 9
Approx. age of heifer: yrs. months	2–6	2–0	2–6	2–6	3–0	3–0	2–6	3–0
Weight (gm.)	2220	1640	2500	2920	4390	6160	9540	6140
Weight as % of weight at first month of pregnancy	100	74	113	132	198	277	430	277
Per cent. increase	—	—	13	19	66	79	153	—

small, but they were animals of similar type and age, and were kept under similar conditions. The individual variation, however, is large (for example, the small weight of the udder at the 2nd month), and although there is increase from the 1st to 3rd month, both weights lie within the normal variation in virgins of the same age. More material is required before a definite growth curve can be drawn. The results, however, indicate (as will be seen from the percentage weights shown in Table XXX) that a normal growth curve is followed very similar to that shown in Diagram II for the uterus and its contents, and that there is no alteration in shape either before or after the 5th month of pregnancy. The difference in weight between the 7th and 8th months was probably caused by the fact that P 6 at the 7th month had the udder fluids withdrawn periodically from the udder, and was producing milk; whereas in P 9 at the 8th month the fluids had never been removed before the animal was killed, and milk had not been formed.

The nature of the secretions which can be obtained from the nipples in the different stages of pregnancy has already been referred to above (p. 116 and Table XXII). Woodman and Hammond[169] have shown (from the heifers mentioned here) that the fluid which can be obtained from the virgin gland is similar in chemical composition and appearance to that which is present up to about the 4th month of pregnancy; this serous, slightly whitish and very fluid secretion, contains small quantities of casein (2·36 per cent.), lactose (1·19 per cent.) and fat (0·12 per cent.), as well as considerable amounts of albumen (1·61 per cent.) and globulin (3·44 per cent.). At the 5th month the character of the secretion changes, and it becomes very viscid, yellowish and honey-like.

Woodman and Hammond[323] have shown (from the heifers mentioned in this paper) that at this stage the secretion consists almost entirely of globulin (34·5 per cent.), with small quantities of albumen, (1·0 per cent.) and only a trace of casein and lactose and fat. They concluded that it is the mixture of this substance with true milk, which

PLATE XXXII

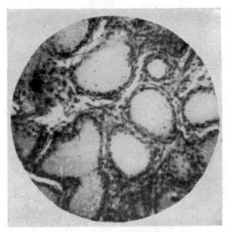 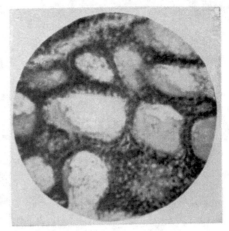

1. Seventh month (fluids withdrawn). 2. Eighth month (fluids not withdrawn).

ALVEOLAR DEVELOPMENT IN THE MAMMARY GLAND
DURING FIRST PREGNANCY (*cont.*).

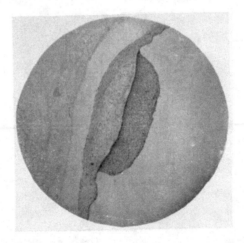

3. Third month of pregnancy.

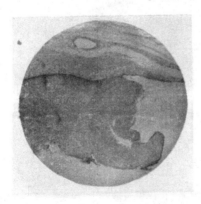 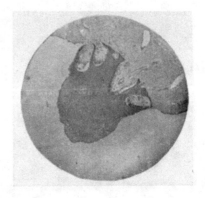

4. Fifth month of pregnancy. 5. Eighth month of pregnancy.

AMNIOTIC PUSTULES.

leaving the foetal membranes and placenta intact, when neither the corpus luteum nor the mammary gland continued to develop.

This evidence is, however, only indirect, for it is based on the assumption that it is the foetus which causes the corpus luteum to persist, and that this in turn acts on the mammary gland. Direct evidence will be difficult to obtain until such a time as it is possible to produce the persistent corpus luteum experimentally.

The majority of injection experiments on the effect of extracts on the growth of the mammary gland such as were performed by Aschner and Grigoriu (336) and Fellner (337) are open to the criticism (see p. 81) that the results obtained are due to the normal variations in ovarian activity, since nearly all have been performed on uncastrated animals. Lane-Claypon and Starling (338) concluded that the foetal secretions were the direct cause of mammary development, but although the discovery of the effect of the corpus luteum has rendered their conclusions doubtful, it has not yet been conclusively shown that the foetus has no direct action. Biedl and Königstein (339) from injection and implantation experiments concluded that the foetus had an action on the growth of the gland, but the placenta was without effect.

Loeb and Hesselberg (334), however, have shown that if guinea-pigs are castrated during pregnancy, no proliferation of the mammary gland occurs, even though the foetus is present; we have not, however, in the rabbit found it possible to remove the ovaries without causing abortion.

The effect of the uterus itself on the primary mammary growth has been eliminated by hysterectomy experiments (Hammond and Marshall (60)), and Craig (17) states that in cases of pathological accumulation of fluids in the uterus no enlargement of the mammary gland occurs.

One reason why the development of the alveoli of the mammary gland is not apparent until the 5th month of pregnancy has already been given (p. 176), but other explanations are possible.

A study of the pregnant uterus and its contents (p. 130) shows that there are three changes occurring at the 5th month of pregnancy which synchronise with the development of the glandular phase of development: (1) the cessation of much further formation of the foetal fluids, (2) the maximum development and breakdown of amniotic pustules, and (3) the completion of cotyledonary attachments occurring just before this time.

Without chronological data for other species or experimental evidence, it is impossible to say whether any of these factors have a direct

H

action on the growth of the mammary gland, or whether both are determined by a common cause—the persistence of the corpus luteum.

The possibility exists that during the first part of pregnancy the secretions of the corpus luteum are utilised to a greater extent in the development of the cotyledonary attachments and fluids than in the latter part, where more is available for mammary growth. Frank[340] concluded from injection experiments that the corpus luteum secretion is stored in the placenta, from whence it is distributed for the use of mammary growth in the second part of pregnancy. The experimental evidence in rabbits is, however, against this view, for the mammary gland does not develop faster during pseudo-pregnancy than in ordinary pregnancy (Hammond[208]).

From the fact that in *Dasyurus* (a marsupial) the whole process of mammary growth and lactation is brought about by the corpus luteum (O'Donoghue[341]) it seems unlikely that in the higher mammals a dual control should have arisen other than control through the duration of the corpus luteum itself.

V.

STERILITY.

No direct investigation of the causes of sterility in cows has been made, but incidentally during the experiments a certain amount of data has been collected which has a bearing on the subject.

The cow is a very suitable subject for the study of sterility, for the ovaries and uterus can be examined per rectum without the necessity of an abdominal operation, and observations on the effect of treatment can be made on the living animal.

It would appear to be of no consequence whether service occurs early or late in the "heat" period (see p. 176), and it may be found possible by certain methods of artificial insemination to prolong the time when conception is possible.

It has generally been recognised that sterility may be either (1) temporary or (2) permanent. The former (1) though not so striking in its effects or so often coming under the notice of veterinary surgeons, is probably the main cause of loss in cow-keeping, both because of the large number of animals so affected and also because of the loss of time it involves in milk production. Albrechtsen[120] gives statistics to support this statement.

The latter (2) is of more importance in pedigree animals, for if developed in commercial stock, it is generally more profitable to slaughter than to treat. Even in pedigree animals, it is probably advisable to slaughter in the case of heifers with what may be called "genetic" sterility; for, as shown by the Duchess tribe of Shorthorns, it is inherited. When, however, a cow becomes sterile in after-life it is probably the result of accidental causes which are not likely to be inherited, and if cured would save a valuable breeding animal.

There is, however, no hard and fast line between temporary and permanent sterility, and below the various causes are described, those of a more temporary nature being taken first.

(a) *Short or faint heat periods*, such as have been described above (p. 18), due to individuality, time of year, shortness of life of the corpus luteum, etc., and which are physiological rather than pathological in origin, frequently cause delay in service rather than absolute sterility.

The symptoms of heat can generally be increased by keeping the animals warm in cold weather, and by giving exercise. Extreme fatness is inimical to intense heat periods; the maximum reproductive functions occur with a rising condition of nutrition rather than with a stationary or lowering condition of nutrition.

Reinhardt (34) suggests the use of yohimbine or cantharides in these cases; but we have not been able to increase the length of the heat period by their use (see p. 20).

(b) *Over-excitability of the animal when on heat* often results in severe straining after service, and frequently extrusion of the semen. Such cases in mares have been treated successfully by artificial insemination (Marshall and Crosland (342)). In cows, where this operation is difficult on account of the small size of the cervix, a remedy sometimes used to prevent straining after service is to throw a pail of cold water over the animal, and also to keep it away from others, shut up in a box by itself; it has then no opportunity to jump other cows, for this is usually followed by straining, and a flow of mucus from the vulva. Reinhardt (34) suggests the use of bromide or other drugs in such cases.

(c) *Impermeability of the cervix to spermatozoa*, caused by the incomplete liquefaction of the mucus of the cervix. As shown above (p. 54), liquefaction normally occurs just before the onset of heat, and the fluid state of the mucus allows the spermatozoa freedom of movement.

Macomber (343) found that in women a thick tenacious secretion of the cervix often causes sterility. It is probable that this condition is aggravated by an acid condition of the reproductive tract, for mucus becomes tenacious in an acid medium, but liquefies easily in dilute alkalis; hence the good results which have been frequently obtained by washing out the genital passages with dilute alkalis just before service. Reinhardt (34) suggests the use of a 0·5 per cent. solution of sodium bicarbonate.

The thickened condition of the mucus of the os is probably one of the causes why forced service when an animal is not on heat (see Lewis (344) in pigs) is generally without result; although it is said by Sakowsky (345) that cows can become pregnant by artificial insemination at a time when they are not on heat. Difficulties in getting through the cervix during the interoestral period might be overcome by injections with a covered needle syringe such as is used for the withdrawal of fluids from cysts.

We have frequently seen cases of heifers and cows which come on heat quite regularly, and in which the organs were normal, anatomically, but which failed to conceive after service (see P 8); we believe that many of these cases could be grouped under (b) and (c) above.

Heifer P 8. This heifer was selected for experimental purposes, and the times of heat determined (Tables III and IV); she was mated (together with three others which proved to be successfully in calf) with the bull in December 1921, but did not conceive. She was then taken to another farm, where she was subsequently mated several times with different bulls, but failed to breed. She was killed in July 1922, and the organs were found to be normal in every respect; a fresh corpus luteum was found in one ovary, and the uterine mucosa was congested (see Table XXXII).

Table **XXXII.** *Weights (gm.) and sizes (cm.) of parts of reproductive organs of sterile cows.*

					Ovaries					
			Weight		Corpora lutea size		Follicles or cysts size		Uterus weight	Mammary gland weight
No. of cow		Condition of cow	Largest	Smallest	New	Old	Largest	Next largest		
P 8	Heifer	Organs normal	4·5	4·3	1·1	1·1 0·5	0·9	0·9 0·8 0·7	240	4900
13	Cow	Persistent corpus luteum	16·0	3·5	2·7	—	1·4	—	280	2500
29	,,	,, ,,	14·6	6·5	2·0	0·6	1·8	—	600	2680
10	,,	Large single cyst	26·2	11·1	1·1	—	3·1	—	560	6620
28	,,	Multiple small cysts	12·5	12·1	—	—	2·0	1·4 1·3 1·2 1·1	740	6740
9	,,	Tubercular	6·0	4·0	0·5	—	0·9	—	360	2360

A possible explanation of these cases is that "heat" (desire) and ovulation are not properly correlated, so that service occurs at a time unsuitable for the fertilisation of the egg. There are, however, no data to support this hypothesis, but it could easily be determined by examination of the ovaries per rectum. Strodthoff[346] found in the cow cases of heat without ovulation occurring; but in these cases cysts usually form in the ovaries.

(*d*) *Pathological conditions of the uterine and cervical mucosa* have frequently been cited as causes of sterility. Albrechtsen[120], who believes this to be the main cause of sterility and to be contracted very largely by infection at parturition, has obtained good results in these cases from uterine irrigation. Whether as stated by him uterine infection causes persistent corpora lutea, and so is the primary cause of sterility, we doubt; for many animals with uterine infection come on heat regularly; it is more probable that a persistent corpus luteum aggravates infection by raising the nutritive conditions of the uterus. Beaver, Boyd and Fitch[347] found when making a bacteriological examination of the organs of sterile cows that in most cases there was multiple follicular cystic degeneration of the ovaries, and that this was frequently associated with hypertrophy of the cervical folds.

Many of the cases of inflamed cervix quoted by Albrechtsen were also known to have cystic ovaries, and we believe that the latter is the cause rather than the result of the inflamed cervix; for normally cervical congestion occurs just after each heat period (see p. 54). He states definitely, however, that diseased ovaries and nymphomania can be cured by uterine irrigations.

Albrechtsen's method of manipulation is to hook the cervix and to draw it back towards the vulva; after the insertion of a wire vaginascope, the cervix can be inspected, and, if necessary, a catheter can be introduced into its orifice with the left hand (which also holds the hook) and fluid is pumped in by an assistant; the removal of the fluid from the uterus is effected by pressure of the right hand on the uterus per rectum. Illustrations of the apparatus used are shown in his book. We have found that the insertion of even a small catheter into the cow's cervix is a difficult matter, and entrance is probably much better attained by high pressure with fluids than by the use of dilating instruments.

(*e*) *Cervical stenosis*, or fibrous condition of the cervix, is, we believe, very rarely a primary cause of sterility. Extreme rigidity of the cervix is normal in cows. Pathological growths are sometimes found in the cervix (Albrechtsen [120]), but in our opinion play but a small part in causing sterility. Williams [286] found only one case of closure of the cervix in 1500 animals examined.

Cysts in the external os and vagina often exist (see Oppermann [239]), but they seldom completely block the passage. Oppermann and Keller [348] have also illustrated cases in which the cervix is double (as occurs in the rabbit), but these are rare; the latter shows that the ovaries are functional in these cases, for corpora lutea were found in them.

In our opinion, as is pointed out by Albrechtsen, occlusion of the lumen of the cervix is much more likely to occur by swelling of the mucosa as the result of congestion than by fibrosis of its walls. Treatment in such cases depends on reducing the congestion rather than by dilating the cervix.

(*f*) *The Graafian follicles may fail to mature properly*, and the heat periods may not occur. Whether this condition is permanent or temporary depends on the cause of the failure to develop. Extreme (permanent) cases are to be seen in freemartins (which occur in about 95 per cent. of heifers born twin to a bull) where the ovaries are rudimentary or absent.

Follicles fail to develop in the ovaries before puberty, in the anoestrous condition outside the breeding season, and they also sometimes fail to develop during the breeding season under conditions of reduced nutrition brought about by low feeding during periods of heavy milking (rabbits—see Marshall and Hammond(24)) or cold wet weather.

In heifers delay in coming on heat has been noticed which could be attributed to this cause; but only one case of complete absence (C 1) due to food and weather, and one to disease (A 4) have been observed. In another cow affected with tuberculosis the ovaries contained neither recent corpora lutea nor large follicles (Cow No. 9), but the breeding history of this animal was unknown.

Heifer C 1 (see Tables III and IV) was in poor condition during December, and was due to come on heat on December 28th. She was watched night and morning every day from the time of her last heat up to December 25th, and also from January 1st to 15th; from the morning of December 26th to December 31st she was tried with the bull every two hours, but showed no signs of heat. On January 16th she came on heat, but the symptoms were slight. Previously her heat periods had lasted 18–20 hours. No bleeding occurred from the vulva between December 11th and January 16th, although she had previously bled after her heat periods, and it is fairly certain that no heat occurred between December 10th and January 15th.

Cow No. 9. From a slaughterhouse; carcase condemned for generalised tuberculosis. Ovaries (Plate XX, 3 and Table XXXII) contained neither recent corpora lutea nor follicles of any size, the largest corpus luteum being only 0·5 cm. in diameter and the largest follicle 0·9 cm. Unfortunately no previous history of the cow was obtained.

Cow A 4. Experimental animal; 12 years old; dry and in good condition. Purchased in January and came on heat regularly until the end of March; remained on heat about 17 hours (see Tables III and IV). She was due to come on heat on April 12th and again about May 1st, but although seen at least twice daily (8 a.m. and 4 p.m.) from April 3rd to April 8th, and tried with the bull every two hours from April 9th to May 4th, when she was killed, no signs of heat were observed. At slaughter, the carcase was condemned for generalised tuberculosis; the ovaries (Plate III, 3) showed that ovulation had not occurred for some time, as only a very old corpus luteum, scarlet in colour and of 0·7 cm. diameter (Table XXXIII) was present in the ovaries. In the left ovary there was a follicle of almost mature size (1·1 cm.), but it did not project from the surface of the ovary as most ripe follicles do; it showed in the liquor folliculi on the inner side of the follicle a large clot of blood. A similar extravasation of blood into one of the large follicles (0·95 cm.) of the right ovary had also occurred.

Atrophy of the follicle associated with extravasation of blood into the cavity, as seen in Cow A 4, has also been observed even in small follicles of 0·5 cm. diameter, after the ovary had been squeezed in the hand per rectum (Cow A 2). It should be stated that the ovaries of

Cow A 4 were not handled, nor was any vaginal or rectal examination made during the time she was in our possession, from January to May. The microscopic appearance of the atrophic follicle in Cow A 4 is shown in Plate XXVI, 2; the granulosa of the follicle appears to be intact and normal; but some of the blood vessels of the theca have broken down, and red blood corpuscles have penetrated the granulosa and lie in the liquor folliculi. Since the occurrence of degenerate follicles of this type is much more frequent in the rabbit (where spontaneous ovulation does not occur) than in other animals, it would appear to result from congestion of the theca which was not followed immediately by rupture of the follicle. The relation in this case to the tubercular condition of the cow is not clear, unless it be that supplies of generative ferment are reduced by tuberculosis.

In a similar but small follicle from Cow A 2, in which the ovary had been squeezed five days before the animal was killed, the blood half filled the follicular cavity, and the granulosa cells were degenerate and scattered about the cavity; a few had developed into small luteal cells. Fitch (349) states that he has occasionally found small haemorrhages in the walls of cysts, and we have also observed blood corpuscles percolating through the granulosa of cysts (Cow 10).

(*g*) *Persistence of the corpus luteum.* Only two cases have been seen in which sterility could be attributed to persistence of the corpus luteum.

Cow No. 13 was obtained from a slaughterhouse, and its previous history traced. It was about 4 years old, and was sold in October because it was barren; it was put up to fatten, and killed in the following March. The ovaries (Plate XX, 3 and Table XXXII) contained a corpus luteum which was larger (2·7 cm.) than any found during the cycle or during pregnancy. A follicle of mature size (1·4 cm.) was also present in the ovaries; since in the normal cycle large follicles and corpora lutea of the largest size do not occur together it was concluded that this abnormality was the cause of sterility.

Cow No. 29 was obtained from a slaughterhouse and was known to have been killed because she was sterile. The ovaries (Plate XX, 3) contained a corpus luteum of full size (2·0 cm.) together with a follicle which was considerably larger than the normal (1·8 cm.) (see Table XXXII).

Herdsmen have informed me that some cows regularly go 4–5 weeks between their heat periods, and that such cases are difficult to get in calf.

The general symptoms of persistent corpora lutea are long inter-oestral periods, *i.e.* the corpus luteum persists as it does in pregnancy, and does not atrophy as usual after a life of three weeks.

The cause of persistent corpora lutea Zschokke (350) states is due to several reasons; normally it is caused by the development of the foetus in the uterus and pathologically he believes that it can be produced by retained after-birth and uterine infections.

Bongardt[351] found that after contagious abortion cows are very liable not to come on heat for some time. Strodthoff[346] states that uterine infection must be well developed before the corpus luteum persists, but that after healing of the uterine mucosa, the persistent corpus luteum atrophies. Albrechtsen[120] too bases his treatment of persistent corpora lutea on washing out the uterus. Many of these cases are, however, open to another explanation (foetal atrophy, see (*i*) below).

It appears to us uncertain whether infection of the uterus in itself causes the persistent corpora lutea, or whether a persistent corpus luteum intensifies the uterine infection. We have not been able by operations on the uterus of the rabbit to make the corpus luteum persist in the ovaries (see p. 126).

Zschokke[350] also states that persistent corpora lutea are produced by the use of certain foods—rye, malt, brewers' grains and sugar beet silage. He did not, however, obtain experimental evidence for this statement, but based it on experience in practice. It is known that these foods increase milk secretion, and two possibilities exist to account for the non-occurrence of heat in such cases: (1) that the increased milk secretion prevents the ripening of follicles as it does in lactating rabbits, and so gives rise to the same symptoms as persistent corpora lutea, *i.e.* absence of heat periods; or (2) that these foods contain specific substances especially suitable for the nutrition both of the corpus luteum and of the mammary gland.

Schmid[16] gives the weights of the ovaries and sizes of the corpora lutea of a number of cows in which heat had not occurred for some time (3–8 months). In the majority of cases the persistent corpora lutea were of much the same size as those of the cycle at the stage of maximum development; they were also identical in histological structure. Kalte-negger[186] describes the histological structure of a number of persistent corpora lutea in the cow; he states that he could observe no essential difference between it and the corpus luteum of pregnancy, or the cyclic corpus luteum at the stage of maximum development; the two former differ from the latter only in that degenerative changes in the luteal cells and overgrowth of connective tissue do not appear so soon.

The treatment of cases of sterility due to persistent corpora lutea has been practised for some time with considerable success in Switzerland and elsewhere by Hess[41] and others. Hess gives a well-illustrated account of the operation of squeezing out the persistent corpora lutea from the ovaries per rectum or per vaginam; he states that corpora lutea up to 5 months old can easily be squeezed out, but that after this time it is more difficult as the corpus luteum tends to sink into the centre of the ovary.

He states that in 50 per cent. of cases the cows come on heat again on the evening of the 3rd or morning of the 4th day after the operation; in 20 per cent. heat occurs from the 4th–10th day after the operation, and in 10 per cent. between 10 and 28 days afterwards; he found that 95 per cent. became pregnant at the first heat period if served. Schumann[352] was also successful with this operation, only 8 out of 70 cases treated failing to come on heat.

Hess[41] also describes hypertrophied although not persistent corpora lutea after infection with granular vaginitis and states that in these cases the heat periods are often abnormally long, but they occur regularly. Reinhardt[34] states that the hypertrophied corpora lutea can be caused to atrophy by the use of drugs—Pulv. Myrrhae or Copaiva balsam and Ol. Terebinth.

Numerous cases of corpora lutea cysts have been described by Fitch[349] and others; but, although some of these may have been pathological, we are inclined to believe that the majority represent a normal early stage in the formation of the corpus luteum (see p. 39). Both Fraenkel (L.)[353] and Fraenkel (E.)[354] have described luteal cysts in women, and the former states that three types can be found, with the inner layer (1) of luteal tissue, (2) of connective tissue and (3) of connective tissue lined by epi- or endothelium. It appears to us probable that the first is a normal stage in the formation of the corpus luteum, in which the follicular cavity is re-distended with fluid, and that the other two are later stages of the same condition, in which for some reason the connective tissue growth in the centre of the corpus luteum has taken place without associated blood vessels to carry off the fluids collected. We have sometimes seen small spaces in the central connective tissue plug of the corpus luteum, and similar cases have been described by Schmid[16], Küpfer[89], Delestre[100] and others in corpora lutea which could not have been in an early stage of development.

Although cystic corpora lutea have been described by many investigators, it has not been shown that they are associated with sterility or any derangement of the cycle or of pregnancy.

(*h*) *The occurrence of ovarian cysts.* The most frequent pathological condition of the ovary was found to be due to the presence of follicular cysts. We agree with Simon[109], who examined a large number of

PLATE XXXIII

1. Wall of large follicular cyst.

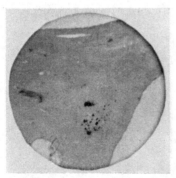

2. Wall of small cyst.

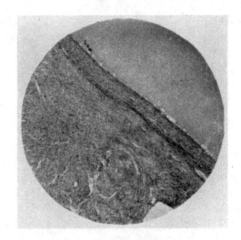

3. Site of corpus luteum which has been
squeezed out, and wall of large unruptured follicle (above).

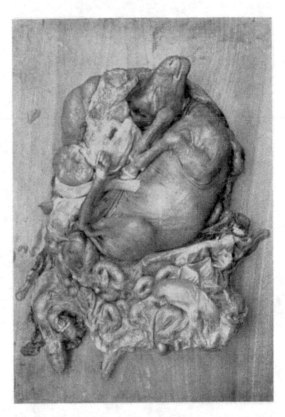

4. Early stage in formation of a hippomane (sheep).

pathological ovaries, and found that in the majority of cases small multiple cysts were present, and that in a few cases large single cysts occurred. The following is a list of the cases we have examined:

Cow No. 10 was obtained from a slaughterhouse and the previous history traced. She calved last on 10. iii. 19 and came on heat on 4. ix. 19, when she was served but failed to conceive. She subsequently came on heat on 22. ix. 19, 15. x. 19 and afterwards every fortnight; although frequently served she never became pregnant, and so was put up to fatten and was killed in the following March. The ovaries (Plate XX, 3) showed a large follicular cyst, 3·1 cm. in diameter (Table XXXII), nearly three times the size of an ordinary ripe follicle. This contained liquor folliculi of fairly thick consistency, and sections showed that the granulosa was intact and normal, and that the theca interna had fairly numerous blood vessels; a number of red blood corpuscles had penetrated the granulosa in certain parts, and were lying between the granulosa and liquor folliculi (Plate XXXIII, 1).

Cow No. 28 was the property of a neighbouring farmer, and produced her first calf in September 1917; she did not calve again until October 1919, but by the end of November 1919 she became dry, and from this time until April 1920, when she was killed, she came on heat regularly, and was frequently served by the bull, but never became pregnant. Her ovaries (Plate XX, 3) showed a number of cystic follicles, the largest (Table XXXII) being 2·0 cm. in diameter, much larger than the normal ripe follicle; many other cysts about the size of a normal mature follicle were present (1·4, 1·3, 1·2 and 1·1 cm. in diameter). No corpora lutea were seen in the ovaries, and although she had come on heat, she had not ovulated for some considerable time.

Histological preparations showed that the largest cyst was without a granulosa layer, its wall being lined by the connective tissue of the theca (Plate XXXIII, 2). The other smaller cysts, however, had quite normal granulosa layers.

Another pedigree cow came on heat regularly for over a year, but did not conceive, although served by the bull several times. When killed, both ovaries showed multiple follicular cysts, similar to those described in Cow 28; no corpora lutea were present in either ovary; and although she had come on heat, she had evidently not ovulated for some considerable time.

The cause of sterility in these cases can be attributed to the fact (as shown by the absence of corpora lutea and presence of cysts) that ovulation did not occur, although the animals came on heat. The ripe follicles, instead of rupturing after heat, appeared to go on developing until a large size was reached (Cow No. 10) or, more frequently, they remained about the size of a normal ripe follicle, and others in turn developed, but did not rupture. Similar cases have been described by Pugh [355] and Lothe [356].

It is presumed that the maturation of each follicle was associated with a heat period, but no definite times of heat have been obtained by direct observation of animals with cystic ovaries. From these cases it would appear that heat (desire for coitus) depended ultimately on the maturation of the follicle.

Since in cows with cystic ovaries no corpora lutea are formed to regulate the growth of the follicles, it frequently happens that the ripening of the follicles proceeds unchecked, and the animal becomes a nymphomaniac, coming on heat continuously or at very short intervals (like a rabbit in the breeding season). The symptoms of nymphomania, besides constant "bulling," are swelling of the vulva, a high tail head and sinking of the pelvic ligaments (Hess (41)). Oppermann (239) gives an illustration of the external genitals of a cow with an ovarian cyst which shows a swollen vulva and sinking of the pelvic ligaments, changes very similar to those which occur just before calving, and which have been described by Fuhrimann (250).

Since either one large or many small cysts are usually formed, it would appear that a large cyst by its growth utilises the nutriments required for the growth of follicles, and prevents their development just as the corpus luteum does. If the cystic follicle, however, does not increase much in size, development of other follicles will take place, and multiple small cysts are formed. It is known that in a certain small percentage of cases of cows with cystic ovaries, heat does not occur at all. Schumann (352) states that in cows with cystic ovaries 70 per cent. are nymphomaniac and 30 per cent. are without heat periods. Whether absence of heat occurs in all cases of the formation of single large cysts and nymphomania in the cases of multiple small cysts, is not certain; but would appear probable. In the case of Cow No. 10 above, where a large cyst was present, she had unfortunately not been tried with the bull for some time before killing, as she had been put up to fatten. The following case is, probably, one which could be explained on this hypothesis:

A pedigree Shorthorn cow which had not been seen on heat for nearly two years was found by rectal examination to have a large cyst in the left ovary. This was punctured from the rectum, and the fluids withdrawn by a covered needle cyst syringe. She came on heat again about 14 days afterwards.

Pearl and Surface (167) have described a case of small multiple cysts in the ovaries of a cow in which no corpora lutea were present in the ovaries; they concluded that the symptoms exhibited by the cow in jumping others implied an assumption of secondary male characters, but this we believe is also a symptom of nymphomania.

With cases of multiple small cysts in the ovary, it is often a difficult matter to decide which are cysts and which normal follicles. Rubeli (quoted by Fitch) states that in the cow any follicle over 1·5 cm. in diameter is cystic, for no ova could be found in them. Fitch (349) found in the cow many cysts, not exceeding 1·0 cm. in diameter, and has observed them up to 15·0 cm. diameter; he states that the contents are usually a colourless watery fluid. In our experience the contents of the smaller cysts are very similar to those of the follicles; but sometimes are rather yellow in colour. Hinrichs (357) has observed the yellow colour to be marked in ovarian cysts in mares, and it is not unlikely that, as the granulosa persists, lutein may be passed into the liquor folliculi instead of to the ovum. Hess (41) has given an analysis of the contents of a large cyst (about 100 gm.) which was of yellow colour—water 93·28 per cent., proteins 5·49 and ether extract 0·02 per cent.

Fitch (349) states that the outer wall of the cyst is of connective tissue, and the inner surface is composed of several layers of small round or slightly oval cells which vary in thickness in different places (granulosa?).

In our specimens the very large cyst (Cow 10) had a normal, although thin, granulosa layer, but one of the many cysts in Cow 28 had no granulosa, although the layer was present in the other cysts of this ovary.

There can be no doubt that these were true cysts, in Cow 10 because of the size and in Cow 28 because of the number, and in each case no corpora lutea were to be found in the ovaries, although heat had occurred. The presence of a normal granulosa layer in cysts is surprising, and would seem to indicate that the cause of failure of the follicle to rupture does not rest in this layer. Since in Cow 10 red blood corpuscles had penetrated into the cavity of the cyst, it is presumed that an unsuccessful attempt to rupture had been made, but had been prevented either by fibrosis of the theca or by thickening of the tunica albuginea. In fowls, where the follicle has a stigmen or line of rupture, multilocular cysts may be formed in the ovaries, and Laurie (358) has described cases where bleeding occurred into the cyst cavity.

It has been shown (p. 129) that in follicular degeneration, which is always attended by decrease in size, the granulosa is the first layer to break up; it is not unlikely that if the granulosa remains intact when the follicle fails to rupture, further increase instead of decrease in size takes place. Eventually, no doubt, the granulosa degenerates and disappears, and in Cow 28 the cyst without granulosa was probably of long standing, while those more recently formed still contained a granulosa layer. Bland Sutton (123) has described the appearance of granulosa in the early stages of cyst formation, but it disappeared in the late stages in the ovaries of women.

Albrechtsen(120) states that cysts rarely form in the ovaries shortly after calving, and do not occur until after the time when the first heat period would have been due. Corner(266) has described several cysts in sow's ovaries, in which the granulosa had begun to form luteal tissue. Meyer(359) has figured a human ovarian cyst in which the appearance was very similar to that of Cow No. 10, but he states that in places small luteal cells were being formed, as a result of the extravasation of blood through the granulosa. Cohn(360) has figured the walls of cysts in women's ovaries which show that granulosa cells are present in the early stages but not in the late stages.

Burghardt(361), who gives a summary of the literature, found that in mares' ovaries multiple small cyst degeneration was common; in small cysts he found degenerating granulosa cells, but in the large ones only theca interna. It is well known that the mare's ovary frequently becomes very overgrown with connective tissue, due to the irregularity with which they are bred from, and it may be that this connective tissue growth tends to inhibit free ovulation.

Simon(109) found that the walls of cows' follicular cysts were generally 0·6–1·0 mm. thick and that granulosa cells were present, but they were larger than in normal follicles. He could find no thickening of the albuginea in these cases, but says that sclerotic tissue is formed in the follicle wall itself from the ovarian stroma, and that this prevents rupture. Zschokke(350) concluded that the absence of the corpus luteum may lead to cyst formation by the absence of the necessary congestion stimulus for ovulation; so that the accidental formation of one cyst leads to the formation of others from ripening follicles, which do not receive the stimulus for ovulation from a degenerating corpus luteum (but see p. 77).

Zschokke states that ovarian cysts may occasionally be formed also from Pflüger's tubes, or by colloid degeneration of the cells of the ovary, but these are not common, and rarely cause nymphomania.

A probable reason why ovulation is prevented and cyst formation occurs is shown by the case of the experimental cow A 3:

Cow A 3. This cow had regular heat periods from January to May. At 9.30 a.m. on May 16th a corpus luteum 6 days old was squeezed out of the left ovary, and the cow came on heat again at 2 p.m. on May 18th. Six days after, on May 24th, she was killed, and the ovaries (Plate III, 3) showed that, instead of ovulating at the last heat period, the ripe follicle, which had formed in the left ovary, had persisted, and formed a cyst 1·75 cm. in diameter; another follicle had almost reached mature size (1·3 cm.) (see Table XXXIII).

Table XXXIII. *Details of appearance and size of reproductive organs of experimental cows.*

No. of cow	History	Ovary	Ovary weight (gm.)	Corpus luteum (size cm.) New	Old	Squeezed out	Follicle (size cm.) Largest	Next largest	Uterus weight (gm.)	Post mortem appearance of organs — Ovaries	Uterus	Vagina etc.
A 2	On heat 27 Apr. At 10 a.m. on 16 May.—Unsuccessful attempt to puncture follicle in left ovary. Ovaries squeezed. On heat 6 p.m. 20 May. Killed 12 p.m. 2. May	Right	14·4	—	0·7	—	1·3	1·0	1320	3 large follicles	Cotyledons rather congested, lower ones more than upper	Very congested. Os dilated and lower fluid mucus with yellow patches in it
		Left	13·1	—	—	Partly squeezed out 1·1 in ovary 1·4 outside ovary	0·6	0·5		Congestion where corpus luteum half squeezed out. Small black follicle near it		
A 3	On heat 15 May. Corpus luteum squeezed out of left ovary 9.30 a.m. 16 May. On heat 2 p.m. 18 May. Killed 5.30 a.m. 24 May	Right	7·3	—	0·5	—	1·3	0·5	640	Large follicle	Both cotyledons and general surface pale	Very congested above urethra. Slightly congested in other parts. Squeezed out corpus luteum* (wt. 1·2 gm. and almond shaped) found free in body cavity between rumen and diaphragm
		Left	9·7	—	—	1·5 thick diam. (almond shaped)	1·75	0·5		Small scar on surface where corpus luteum squeezed out next the large follicle		
A 4	On heat 25 Mar. No heat on Apr. 12 or May 1. Killed 4 May	Right	6·7	—	0·7	—	0·95	—	740	Large follicle	Black pigment on some of cotyledons	Normal; cervix closed with very little mucus
		Left	5·0	—	—	—	1·1	—		Large follicle with dark brown patch on one side		
A 5	On heat 2 p.m. June 3. Corpus luteum squeezed out of right ovary 9 a.m. June 10. On heat 10 a.m. June 12. No further heat up to June 25 at 6 p.m. when killed	Right	7·8	—	0·7	1·6 thick diam.	1·1	1·0 0·7	700	One large and several smaller follicles. Star-shaped scar of white connective tissue where corpus luteum squeezed out.	Surface pale. Cotyledons yellowish brown	Pale. Cervix closed with little thick mucus, normal. Squeezed out corpus luteum (wt. 1·1 gm. almond shaped and orange yellow colour) found free in body cavity just above the uterus
		Left	9·7	2·1	—	—	Not sectioned	0·4		Large corpus luteum protruding from ovary		

* Hess (41) has found these squeezed out corpora lutea free in the body cavity 9 months after the operation and still weighing 2 gm.

The reason why normal ovulation did not take place in this case was believed to be the presence of scar tissue on the surface of the ovary in close proximity to the ripening follicle (Plate XXXIII, 3), the granulosa of which was quite normal. That this failure to ovulate does not always occur after the corpus luteum has been squeezed out is shown in the case of Cow A 5 (but see p. 23).

On the basis of this case it is believed that the prevention of ovulation and formation of cysts may occur as the result of any inflammation (of perhaps only temporary duration) on the surface or depths of the ovary which leads to the growth of tough connective tissue. Such inflammation would be especially liable to occur after cases of contagious abortion, delayed cleansing, metritis, or acute indigestion; all of which are liable to cause an inflamed condition of the peritoneum in the vicinity of the ovary, and many of which have been known for some time to be associated with cases of sterility. Stazzi[362] states that after granular vaginitis, metritis is frequent, and either nymphomania or suppression of oestrus frequently occurs. Peters[363] found that sterility was frequent after contagious abortion, and that the ovaries were often infected with bacteria. Macomber[343] found that in women a multiple small cystic state of the ovaries is usually associated with a thickening of the tunica albuginea, and that chronic congestion of the ovaries usually inhibits ovulation.

The treatment of ovarian cysts in the cow is described by Hess[41], who gives illustrations showing how they may be ruptured by pressure with the hand per rectum or per vaginam, or their fluids may be withdrawn by a covered needle syringe. Hess found that rupture is more successful in recent than in long-standing cases where the whole ovary may have become cystic. In cases where a temporary inflammation has subsided, there is more hope of recovery than where inflammation and infection still exists, and the latter may have to be treated several times before a cure is effected.

The removal of both ovaries has been practised largely in mares which are nymphomaniac and have ovarian cysts (Hobday[364]), but this in the cow would prevent further breeding; although it is said that if the operation is performed when the cow is in full milk, the flow is maintained for a long period (see Hobday[364] and Reisinger[365]).

It may be found useful in cows, where only one ovary is cystic, to remove it, leaving the other intact; the presence of corpora lutea or cysts in one ovary has a regulating action over the other ovary, and if the diseased one is removed, it would allow the other to function normally. Peters(363) found in 74 sterile cows that were slaughtered 42 had only one ovary diseased, while 32 had both ovaries affected.

(*i*) *Foetal atrophy.* Cases of sterility are known to occur where the cow is served by a bull and then, after showing no sign of heat for several months, suddenly comes on heat again, although no abortion has been seen. Such cases are frequent after outbreaks of contagious abortion (see Williams(285)) or granular vaginitis, and are believed by some to be due to persistent corpora lutea, or to unobserved abortion; it is more probable, however, that in these cases foetal atrophy has taken place, and absorption *in situ* has occurred (see p. 137).

Death of the foetus can also occur in the absence of any infection (279).

Foetal atrophy in animals producing several young at birth only causes reduced fertility; but in the cow sterility is the result, and the presence of the atrophic foetus is not realised, because it is absorbed instead of being aborted. A case where an atrophic cow foetus was aborted has been given by Hammond(279). Two cases of foetal atrophy in one foetus of a twin pregnancy in the cow have been reported by Küpfer(89). Sterility in these cases did not occur, for the normal foetus remaining stimulated the corpus luteum, so that pregnancy was continued. Albrechtsen(120) states that in several cases he has observed degenerate foetal tissues on irrigating the uterus of sterile cows. Hess(41) states that in cases of mummified retained foetuses, he has obtained expulsion by squeezing out the corpus luteum, when the uterus contracts, the cervix dilates, and the uterine contents are expelled. In the cow, the squeezing out of the corpus luteum is a simple method of causing artificial abortion.

The comparative incidence of the causes of sterility. Zschokke(350) classifies the causes of sterility in the cow as follows:

(1) When heat is absent—due to
    (*a*) General causes—feeding, heavy lactation, general diseases, etc.
    (*b*) Special causes—freemartins, inbreeding, persistent corpora lutea, tuberculosis, etc.
(2) When heat is feeble—due to heavy milking, disease, keeping shut up, malnutrition, etc.
(3) When heat occurs but conception does not follow—due to
    (*a*) When heat regular—mechanical hindrances, disturbed state after heat (young heifers), full blooded animals straining after service, acidity of vagina, diseases of uterus, etc.
    (*b*) Heat too frequent—nymphomania, diseases of ovary, cysts.

(4) When the animal becomes pregnant, but the foetus perishes, due to—
abnormal development of foetus, abortion and other infections, bad
feeding, etc.

We have questioned herdsmen in order to find out the most frequent
causes of sterility, and their replies (see Appendix I (3)) show that a
very small percentage of cases can be attributed to the absence of heat
periods (*i.e.* persistent corpora lutea or unsuitable conditions for ripening
of the follicle); the majority of sterile cows in this country come on heat
but do not conceive after service, and these would be included in
Zschokke's classes (3) and (4).

Albrechtsen[120] gives the following incidence of cases of sterility:
without oestrus 328, with oestrus 313; he found that in about 1000 cases
of sterility treated, only 15 per cent. had cystic ovaries, but that a large
proportion of these were nymphomaniac; the majority of cases of
sterility he believes are due to uterine infection.

Investigators are divided in opinion as to whether most cases of
sterility are caused by uterine infection, or to ovarian disease.

Scheidegger[366] believes with Albrechtsen that infection of the
uterus and cervix are the main causes of sterility, and treats them by
washing out; he found the following distribution of cases:

Inflammation of vagina 40, cervix 45, ovaries 45, uterus 44, and
Fallopian tubes 27.

Posselt[367] also gives frequencies of the different causes of sterility
in the cow which show that uterine infection (76) is the chief cause,
while ovarian cysts (47) and persistent corpora lutea (43) are very
frequent, and that atrophy of the ovary (12), growths in the cervix
(11) and abnormal vaginal secretions (9) are moderately common
causes.

On the other hand, Hess[41] says that the ovaries are the most frequent seat of sterility, and that cystic degeneration is much more
frequent than persistent corpora lutea. He states that 70–80 per cent.
of cases can be cured by treating the ovaries. Poulsen (see [120]) also
believes that the ovaries are the most common origin of sterility in cows
and more especially persistent corpora lutea.

Schumann[352], who studied the frequency of cases of sterility after
outbreaks of granular vaginitis, found the following distribution of
causes—hypertrophied corpora lutea 63, persistent corpora lutea 60,
ovarian cysts 28, uterine and cervical catarrh 51, pyometra 3, various 16.
Wilson[242], in examining cases of sterility after infections with granular
vaginitis in Essex, found that in 55 per cent. of the animals the ovaries

were affected; he states that the granular lesions in the vagina did not in themselves affect conception to any appreciable extent.

Many investigators state that frequently two or more possible causes coexist and which is the primary and which the secondary cause they are unable to determine. Stålfors(314), who treated a number of cases of sterility in cows, found the following incidence:

	Alone	With ovarian cysts	With persistent corpora lutea	With cysts and persistent corpora lutea
Uterine catarrh	10	12	29	36
Persistent corpora lutea	92	54	—	—
Ovarian cysts	53	—	—	—

He states that he was able to cure 72 per cent. of the cows with persistent corpora lutea and 67 per cent. of those with cysts and quotes many cases to show that the ovum which was fertilised in these cases came from the ovary which had been treated.

SUMMARY.

A study of the changes in the reproductive organs of the cow before puberty, during the oestrous cycle and during pregnancy has been made by weighing and observing the parts.

Statistics show that the breeding season extends throughout the year.

The age of puberty and other facts have been ascertained from herdsmen.

The signs of oestrus are described.

Variations in the length of the oestrous cycle and the duration of oestrus have been studied by the use of a vasectomised bull and the duration of these processes has been timed to within an accuracy of about 2 hours in a number of different animals.

On the whole it was found that the cycle length and duration of oestrus were correlated.

The length of the cycle and the duration of oestrus were greater in the summer than in the winter and longer in thin than in fat animals; individual differences also affected its length. The effect of age was only slight and drugs (yohimbine, camphor), the proximity of the male, and whether coitus occurred or not, had no influence on the length of the cycle.

Removal of the corpus luteum from the ovary performed six days after the previous heat period results in the return of heat within two days, *i.e.* an oestrous cycle of 200 hours instead of a normal one of 450 hours. The duration of the corpus luteum therefore controls the length of the oestrous cycle.

Not only does the removal of the corpus luteum shorten the length of the cycle but it also shortens the length of the subsequent heat period, *i.e.* after the operation a heat period of 11 hours ensues instead of a normal oestrus of 21 hours. This fact is of fundamental importance as regards the cause of oestrus.

The time relation between oestrus, ovulation and menstruation have been determined, the beginning of oestrus being taken as the starting point. Ovulation occurs some time between 30 and 48 hours after the beginning of heat and bleeding from the vulva takes place in most heifers but in few cows 48 to 72 hours after the beginning of heat.

The differences in time relation between oestrus, ovulation and menstruation in the cow, the bitch and in woman are not in accordance with the theory that the ripening of the Graafian follicle is the direct cause of menstruation.

Measurements of the sizes of the follicles and corpora lutea show that the active corpus luteum prevents the ripening of the follicle and it is suggested that this is not caused by the internal secretion of the corpus luteum but by the fact that the corpus luteum probably utilises a substance, existing only in small quantities in the blood, which is necessary for the growth of the follicle.

The early stages of the corpus luteum contain large amounts of liquor folliculi in their centre and, as this disappears with the organisation of the capillaries, it is suggested that its formation supplies the internal secretion of the corpus luteum and that the lipoid which accumulates within the luteal cells as a result of the shedding of the ovum (to which it is normally transferred in the follicle) eventually results in their death.

The changes occurring in the uterus during the cycle are described; it is considered that the cyclic changes are similar in kind but less in degree than those which occur during pseudo-pregnancy (in the rabbit) and pregnancy.

Marked changes in the cervix occur during the cycle; large quantities of fluid mucus are secreted at oestrus but during the remainder of the cycle only a little thick tenacious mucus is present. During pregnancy a large accumulation of thick sticky mucus forms a cervical seal which flows away just before parturition in a similar way to the flow which occurs at oestrus.

There is a marked cycle of changes in the vagina, the structure of which varies in different regions. In heifers the bulk of the bleeding after oestrus is derived from the vaginal mucosa. The appearance of vaginal smears at different stages of the cycle is shown.

The anatomy and histology of the udder is described and also the changes which it undergoes in foetal and post-natal life.

The methods of diagnosing pregnancy are discussed.

During pregnancy the corpus luteum remains large and apparently active up to the 8th month, the latest stage examined; follicular atrophy is described.

Weights have been taken and growth curves plotted of the different parts of the pregnant uterus and its contents; up to the 5th month the foetal fluids have the greatest rate of growth but after this time the foetus itself has the greatest growth rate and the amount of fluids remains constant or declines. The relation of these facts to the cause of birth, abortion or foetal absorption is discussed.

Amniotic pustules increase in size up to the 5th month of pregnancy and then begin to atrophy.

The relation of the foregoing facts to the changes occurring in the mammary glands is discussed, for during pregnancy marked changes (growth of alveoli) in the structure of the mammary gland and in the nature of its secretions occur at the 5th month.

Since bleeding occurs just after fertile service but not later in pregnancy and the attachment of the foetal membranes does not occur until after 30 days from impregnation (the cycle is only 19 days) it is doubtful whether the cyclic bleeding can be a preparation for the attachment of the membranes.

The changes in the muscles and glands of the uterus during pregnancy are described. No myometrial gland was found.

The formation of the placenta was found to be caused by the phagocytic action of the foetal membranes causing proliferation of the subepithelial connective tissue layer of the uterus which is highly developed in the cotyledons. Occasional attachment to the intercotyledonary area also occurs by this means but is usually prevented by the flow of uterine milk from the intercotyledonary glands.

The uterine epithelium of the cotyledon is destroyed by the foetal tissues before attachment is made; the walls of the crypts are not lined by uterine epithelium.

The formation of pigment and hippomanes from red blood corpuscles and uterine milk respectively in the foetal membranes is shown.

The appearance of the foetus at different ages is described and weights and lengths are given.

Various causes of sterility have been investigated and discussed.

APPENDIX I.

Tabulated replies to questions put to experienced stockmen at the Dairy Farmers' (*milk breeds*) and Smithfield Clubs' (*beef breeds*) Shows.

The numbers in brackets represent the number of stockmen who were of this opinion.

Question	Answer		Remarks made
	Dairy Show (22 opinions)	Smithfield (18 opinions)	
(1) At what age do heifers first show signs of "heat"?	Mths. Mths. Guernsey (2) 5 Shorthorn (4) 9 Jersey (3) 8 Devon (2) 11 Friesian (4) 8 Dexter (1) 7 Red Poll (2) 8 Kerry (2) 11 Ayshire (1) 12 Av. all breeds: 9 months	Mths. Mths. Lincoln Red (1) 6 Galloway (1) 12 Aberdeen Angus (2) 6 Kerry (1) 12 Hereford (2) 7 Devon (1) 15 Sussex (2) 7 Shorthorn (1) 12 Welsh (1) 8 Av. all breeds: 9 months	Come on early if well fed, late if badly fed

(2) What difference exists between young heifers and old cows in respect of—*(a)* the length of the cycle?	No difference... (19)
	Heifers longer cycle... (1)
	  ,, shorter cycle... (1)
	  ,, more irregular cycle (1)
		No difference... (11)
		Heifers longer cycle... (1)
		  ,, shorter cycle... (0)
		  ,, more irregular cycle (1)
			One cow which usually had 27 day period. Some with periods of 4–5 weeks, difficult to get them in calf. Extremes, twice a week (barren) and 31 days. Jerseys go 22–23 days, Devons and Friesians 20 days and Red Polls, Shorthorn and Guernsey 21 days.

(b) The duration of heat?	No difference... (13)
	Heifers longer on heat (4)
	  ,, shorter on heat (5)
		No difference... (11)
		Heifers longer on heat (1)
		  ,, shorter on heat (0)
			Stay on heat longer if not served (2). Signs of heat not well marked in Aberdeen Angus and Friesians. Jerseys stay on longer than Shorthorns. Heifers more excited when on heat than cows. Stay on heat longest and seen best in spring and summer. To make go off heat throw cold water over them

Question	Answer: Dairy Show	Answer: Smithfield	Remarks made
(3) On the appearance of blood from the vulva after the heat period			Not half bleed; 85 % bleed. More than half heifers bleed. Very seldom bleed after their first calf. Bleeding due to over-service; if served by bull do not bleed; if bleed are in calf (2); turn to service if bleed too much; breed better if bleed
(a) What is the comparative frequency in heifers and cows?			
More heifers bleed than cows ...	(10)	(5)	
,, cows ,, heifers ...	(0)	(0)	
No difference ...	(8)	(1)	
(b) Has the "condition" of the cow any effect on bleeding?			Individual differences. The "hotter" the animal on heat the more she bleeds
Most blood when fat ...	(4)	(2)	
,, ,, thin ...	(1)	(0)	
No difference ...	(6)	(3)	
(4) What is the time after calving that the first heat period occurs?			If in good condition come on heat in 3 weeks, if in poor condition later (2). Do not come on so soon in winter months (3). If very heavy milkers do not come on so soon (3)
(a) If milked			
Varies from 9 days ...	(4)	(1)	
,, 2 weeks ...	(2)	(0)	
,, 3 ,, ...	(7)	(5)	
,, 4 ,, ...	(5)	(4)	
,, 5 ,, ...	(2)	(0)	
,, 6 ,, ...	(3)	(2)	
,, 8 ,, ...	(1)	(2)	
,, 9 ,, ...	(2)	(0)	
,, 10 ,, ...	(0)	(1)	
,, 3 months ...	(2)	(0)	
,, 4 ,, ...	(1)	(0)	
,, 6 ,, ...	(1)	(0)	
Usual time: 3–4 weeks		3–4 weeks	
(b) If suckling a calf			Mainly old cows which do not come on "heat" when suckling. Feeding mangolds delays coming on heat. When suckling not so regular in their periods. Come on heat but do not hold to the bull. Come on heat when calves removed. Browned oatmeal fed brings cow on heat
Come on longer after calving than if milked ...	(11)	9 { 3 weeks ... (1); 2 months ... (1)	
Come on at same time as if milked ...	(7)	2 { 3 ,, ... (2); 5 ,, ... (1); 6 ,, ... (1)	

(5) Does act of milking make cow go off heat?

No ... ... (14)
Yes ... ... (2)

(6) If a heifer which is in calf for the first time aborts—how long has she to be pregnant before milk is produced?

Months
4 ... ... (1)
5 ... ... (7)
6 ... ... (7)
7 ... ... (5)
8 ... ... (2)
Majority 5–6 months ... 6–7 months

Would increase in flow at proper date of calving. Come to milk as time goes on. A cow which has produced a calf will give milk after a 5 months' abortion but a heifer not until 7 months
(0) (2) (3) (8) (1)

(7) What causes variation in the amount of fluid obtained from the udders of virgin heifers?

Fluid in most heifers (2). Never seen (2). More on grass and with good food (2). More in good heifers. More in animals in good condition. A bad thing, causes gargetty udders (3)

To make udder grow milk well out in first lactation. If udder grows early with first calf will make good milkers. Coffee stimulates milk secretion

(8) What is the proportion of sterile cows in which—

(a) No "heat" periods occur?
None ... ... (5)
Few ... ... (7)
Most ... ... (0)

Come on but do not show it

(b) Come on heat but do not stand to the bull?
None ... ... (0)
Few ... ... (0)
Most ... ... (12)

Generally have irregular periods

(9) Have you yourself known of a case of a cow in calf being served by a bull?

Frequently ... (0)
Occasionally ... (6)
Rarely ... (5)
No ... (1)

Stage of pregnancy when second coitus occurred
3 weeks ... (1)
6 ,, ... (2)
9 ,, ... (1)
3 months ... (3)
5 ,, ... (1)
6 ,, ... (1)
7 ,, ... (1)
8 ,, ... (1)
10 ,, ... (1)

More often occurs in milk than in beef breeds. Took bull and aborted shortly afterwards. Took bull and calved 4 days afterwards

Question	Answer		Remarks made
	Dairy Show	Smithfield	
(10) How long is a cow pregnant?		With heifer calf ... 38 weeks With bull calf ... 41 " Average period ... 40 "	Individuals vary and not with sex of calf (4). Individuals do not vary. Heifers go same time as cows. Big cows go longer than small ones
(11) How soon can diagnosis of pregnancy be made?		By feeling right flank of cow at Months of pregnancy 3 4 5 6 7 (1) (3) (2) (6) (5) Average 6 mths.	Foetus kicks when cow is being milked at 6 months pregnant (2). Nipples sticky when pregnant 3–4 mths. At 4 mths. only if side felt when no food is in stomachs
(12) What are the signs that a cow is about to calve?	(1) Vulva swells (10) (2) Skin drops each side of tail between pin bones (9) (3) Slime from vulva (7) (4) Udder swells (6) (5) Appears uneasy and restless and stands apart from others (4) (6) Passes urine frequently (1)	1st occurs before calving (weeks) 1 2 3 6 (2) (1) (1) (1) Time at which occurs before calving. Few hours (1), 24 hrs. (2). Time before calving 3 hrs. (1), 6 hrs. (1), 14 hrs. (1), 3–4 days (2), 2 mths. traces (1). About one week before (1)	

APPENDIX II.

*Extracts from Logbook. Observations made on a heifer during a
heat period.*

Heifer P 7.

Date	Time	Observations
Nov. 14	4 a.m.	String of fluid mucus from vulva
	12 noon	Fluid mucus from vulva
	4 p.m.	„ „
	6 „	Long string of fluid mucus from vulva
	8 „	Mucus from vulva
	10 „	Mucus from vulva. Bull attempted to serve her but she moved away
	12 midn.	Fluid mucus from vulva
Nov. 15	2 a.m.	Fluid mucus from vulva
	4 „	„ „ „
	6 „	Playing with bull, mucus from vulva
	8 „	Bull tried to jump her but she kept moving away
	10 „	Jumped by heifer P 6 but would not stand to bull when he tried to serve her
	12 noon	Bull served her almost at once, string of clear mucus from vulva after service
	4 p.m.	Bull made several attempts before he served her. Arching back and straining
	6 „	After some delay bull served her. Mucus with tinge of blood in it came away after service
	8 „	Arching back and straining. Bull made several attempts but did not quite manage to serve her
	10 „	Bull served her at once. Yellowish mucus from vulva
	12 midn.	Arching back and straining after service, blood and mucus came away from vulva
Nov. 16	2 a.m.	Arching back and straining. Bull attempted to serve her
	4 „	Bull served her after two or three attempts. Jumped bull. Arched back and straining
	6 „	Bull made several feeble attempts to serve her but she moved away. Little fluid from nipples
	8 „	Bull smelling her but did not attempt to jump her
	10 „	Bull took no notice of her
	12 noon	Bull took no notice of her, yellowish mucus from vulva
	4 p.m.	String of yellowish mucus from vulva
	6 and 8 p.m.	„ „ „
Nov. 17	10 p.m.	String of blood-stained mucus from vulva. [This sponged away so that fresh flow seen]
	12 midn.	Blood-stained mucus from vulva. [This sponged away so that fresh flow seen]
Nov. 18	2 a.m.	Blood-stained mucus from vulva. [This sponged away so that fresh flow seen]
	4 „	A little blood-stained mucus from vulva. [This sponged away so that fresh flow seen]
	6 „	„ „ „ „ „
	8 „	A few drops of blood-stained mucus from vulva. [This sponged away so that fresh flow seen]
	10 „	No mucus or blood
	12 noon	„ „
	2 p.m.	„ „

Heifer C 6.

Date	Time	Observations
June 28	12 midn.	Playing with bull, vulva moist
June 29	2 a.m.	Bull smelling her
	6 ,,	Playing with bull
	8 ,,	Jumped bull. Bull made many attempts to serve her but she kept moving away. Little mucus from vulva
	10 ,,	Jumped bull. Bull served her at first attempt. String of clear mucus from vulva after service
	4 p.m.	Bull made several attempts before he served her. Only a few drops of mucus from vulva after service
	6 ,,	Bull served her at first attempt
	8 ,,	Bull served her at once. Arching back and straining. Thick whitish yellow mucus from vulva
	10 ,,	Bull served her at once
	12 midn.	Thick yellowish white mucus from vulva after service
June 30	2 a.m.	Arched back and straining after service
	4 ,,	Bull not quite so keen, made several attempts to serve her. Arching back and straining. Thick yellowish mucus
	6 ,,	Bull made few attempts to serve her, but she moved away. Yellowish white mucus from vulva
	8 ,,	Bull made several attempts to serve her but she would not stand
	10 ,,	Bull did not attempt to serve her
July 1	4 p.m.	Thick yellow mucous discharge from vulva
July 2	8 a.m.	Blood and mucus from vulva

BIBLIOGRAPHY

(1) SCHMALTZ, *Das Geschlechtsleben der Haussäugetiere*, Berlin, 1921.

(2) KRONACHER, *Allgemeine Tierzucht*, IV. Die Züchtung, Berlin, 1921.

(3) ANCEL and BOUIN, *Archives d. Zool. Expér.* **1** (4th series), 1903.

(4) STEINACH, *Arch. f. Entwick.-Mech.* **42**, 1916.

(5) LIPSCHÜTZ, *The Internal Secretions of the Six Glands*, Cambridge, 1924.

(6) WEBER, *Archiv f. wiss. u. prakt. Tierheilkunde*, **37**, 1911, p. 382.

(7) HEAPE, *Quart. Journ. Micros. Sci.* **44**, 1900.

(8) MAYO-SMITH, *Statistics and Sociology*, **1**, New York, 1895.

(9) GAVIN, *Journ. Roy. Agric. Soc. of England*, **73**, 1912, p. 170.

(10) WILSON, *The Breeding and Feeding of Farm Stock*, London, 1921, p. 94.

(11) TOCHER, *Trans. High. and Agric. Soc. of Scotland*, **31**, 1919, p. 250.

(12) PETERSEN, *Kvaegavl og Kvaegopdraet*, **3**, Odense, Denmark, 1920, p. 164.

(13) HANSEN, *Lehrbuch der Rinderzucht*, Berlin, 1921, p. 438.

(14) HAMMOND and SANDERS, *Journ. Agric. Sci.* **13**, 1923, p. 74.

(15) BONHOTE, *Vigour and Heredity*, London, 1915.

(16) SCHMID, *Inaug. Diss.* Zürich, 1902.

(17) CRAIG, Fleming's *Veterinary Obstetrics*, London, 1912, p. 51.

(18) ELLENBERGER and SCHEUNERT, *Lehrb. der vergleich. Physiol. der Haussäuge-tiere*, Berlin, 1910, p. 705.

(19) WERNER, *Rinderzucht*, Berlin, 1912, p. 531.

(20) HANSEN, *Lehrb. der Rinderzucht*, Berlin, 1921, p. 431.

(21) DECHAMBRE, *Traité de Zootechnie*, III. Les Bovins, Paris, 1922, p. 493.

(22) CUROT, *Fécondation et Stérilité*, Paris, 1921, p. 71.

(23) ECKLES, *Missouri Agri. Exp. Stn. Bull.* **135**; see *Dairy Cattle and Milk Pro-duction* (by same author), New York, 1920, p. 209.

(24) MARSHALL and HAMMOND, *Reproduction in the Rabbit*, Edinburgh, 1925.

(25) KÜPFER, *Vierteljahrsschr. d. Natur. Gesellschaft*, Zürich, **65**, 1920, p. 377.

(26) ZEITZSCHMANN, *Archiv f. Gynaek.* **115**, 1921–22, p. 201.

(27) STRUVE, *Deutsche Landwirt. Tierzucht*, **10**, No. 26, 1906, p. 303 and Fühling's *Landwirt. Zeitung*, **60**, No. 24, 1911, p. 833.

(28) WALLACE, quoted from Marshall, *The Physiology of Reproduction*, London, 1922, p. 43.

(29) MATHEWS DUNCAN, *British Medical Journal*, 1883 (1).

(30) STOCKARD and PAPANICOLAOU, *Biol. Bull.* **37**, 1919, p. 222 and *Amer. Journ. of Anatomy*, **22**, 1917, p. 225.

(31) PAPANICOLAOU and STOCKARD, *Proc. Soc. Exp. Biol. and Med.* **17**, No. 7, 1920.

(32) LONG and EVANS, *Memoirs, University of California*, **6**, 1922, p. 46.

(33) LOEB, *Biol. Bull.* **33**, 1917.

(34) REINHARDT, *Monatsh. für prakt. Tierheilkunde*, **25**, 1914, p. 529.

(35) STÜNKEL, *Diss.* Hanover, 1912.

(36) DUBREUIL and REGAUD, *Compt. rend. Soc. de Biol.* **66**, 1919, p. 139.

(37) LONG and EVANS, *Memoirs, University of California*, **6**, 1922, p. 82.

(38) STOCKARD and PAPANICOLAOU, *Amer. Journ. of Anatomy*, **22**, 1917.

(39) GIRKOVITCH and FERRY, *Compt. rend. Soc. de Biol.* **72**, 1912, p. 624.

(40) ZSCHOKKE, *Landw. Jahrb. der Schweiz*, **12**, 1898, p. 252.

(41) HESS, *Die Sterilität des Rindes*, 2nd ed. by Joss, Hanover, 1921, p. 40.

(42) FRAENKEL, *Archiv f. Gynaek.* **68**, 1903 and **91**, 1910.

(43) LOEB, *Zentbl. f. Physiol.* **23**, 1909, p. 73 and *Biol. Bull.* **27**, 1914.

(44) RUGE, *Archiv f. Gynaek.* **100**, 1913, p. 20.

(45) SEITZ, *Archiv f. Gynaek.* **115**, 1921–22.

(46) CORNER and HURNI, *Amer. Journ. of Physiol.* **46**, 1918, p. 483.

(47) LOEB, *Science*, **48**, No. 1237, 1918, p. 273.

(48) KÜPFER, *Denkschrift der Schweiz. Natur. Gesel.* **56**, 1920.

(49) BIEDL, *The Internal Secretory Organs*, London, 1913, p. 409.

(50) HEAPE, *Proc. Roy. Soc.* B, **76**, 1905.

(51) ROBINSON, *Trans. Roy. Soc. of Edinburgh*, **52**, 1918, p. 303.

(52) PEARL, *Animal Husbandry Investigations Report*, Maine Agr. Exp. Stn. No. 92, 1915.

(53) MARSHALL and PEEL, *Journ. of Agric. Science*, **3**, 1910, p. 383.

(54) MARSHALL, *Phil. Trans. Roy. Soc.* B, **196**, 1903.

(55) MARSHALL, *Physiology of Farm Animals*, Cambridge, 1920.

(56) FRANCK-ALBRECHT, *Tierärztlichen Geburtshilfe*, Berlin, 1914.

(57) LOEB, *Biol. Bull.* **26**, 1914.

(58) MARSHALL and RUNCIMAN, *Journ. of Physiol.* **49**, 1914.

(59) STRODTHOFF, *Arch. f. Tierheilk.* **48**, 1922, p. 28.

(60) HAMMOND and MARSHALL, *Proc. Roy. Soc.* B. **87**, 1914.

(61) HALBAN and KÖHLER, *Archiv f. Gynaek.* **103**, 1914.

(62) WEYMEERSCH, *Journ. d'Anat. et de Physiol.* **47**, 1911.

(63) ANCEL and VILLEMIN, *Compt. rend. Soc. de Biol.* **63**, 1907.

(64) SCHRÖDER, *Archiv f. Gynaek.* **101**, 1913–14 and **104**, 1915.

(65) MILLER, *Archiv f. Gynaek.* **101**, 1913–14.

(66) MARCOTTY, *Archiv f. Gynaek.* **103**, 1914.

(67) LEOPOLD and RAVANO, *Archiv f. Gynaek.* **83**, 1907.

(68) SIEGEL, *Münch. Med. Wochensch.* **63**, 1916, p. 748.

(69) NOVAK, *Biologisches Zentbl.* **41**, 1921, p. 1

(70) TRIEPEL, *Anat. Anz.* **46**, 1914, **48**, 1915 and **52**, 1919.

(71) HEAPE, *Roy. Soc. Phil. Trans.* B, **188**, 1897.

(72) MEYER, *Archiv f. Gynaek.* **113**, 1920.

(73) MARSHALL and HALNAN, *Roy. Proc. Soc.* B, **89**, 1917.

(74) CROOM, *Trans. Ednb. Obstr. Soc.* **21**, 1896.

(75) LOEB, *Zentralblatt f. Physiol.* **25**, No. 9, 1911.

(76) MARSHALL, *Physiol. Reviews*, **3**, 1923, p. 335 and MARSHALL and WOOD, *Journ. Physiol.* **58**, 1923.

(77) HEAPE, *Quart. Journ. Microscop. Sci.* **44**, 1900.

(78) MARSHALL, *Quart. Journ. Microscop. Sci.* **48**, 1904.

(79) MARSHALL and JOLLY, *Roy. Soc. Phil. Trans.* B, **198**, 1905.

(80) BUSQUET, *La Fonction Sexuelle*, Paris, 1910.

(81) WEBER, *Milchwirtsch. Zentbl.* 1911, p. 1.

(82) WEBER, *Deut. Tierärztl. Wochensch.* 1910, p. 671.

(83) SCHMALTZ, in Ellenberger's *Handb. d. Vergleich. Mikros. Anat. der Haustiere,* II, Berlin, 1911.

(84) WILLIAMS, *Veterinary Obstetrics,* Ithaca (New York), 1917.

(85) SISSON, *Anatomy of the Domestic Animals,* London, 1914.

(86) MONTANÉ and BOURDELLE, *Anatomie régionale des Animaux Domestiques,* II, Paris, 1917.

(87) IVANOFF, *Arch. des Sci. Biol.* **12**, 1906.

(88) BOUIN and ANCEL, *Compt. rend. Soc. de Biol.* **67**, 1909, p. 464.

(89) KÜPFER, *Denkschr. Schweiz. Natur. Gesel.* **56**, 1920.

(90) NIELSEN, *Aarsskrift, Kgl. Vet. og Landbohøjsk.* Copenhagen, 1921.

(91) KRUPSKI, *Schweiz. Arch. f. Tierheilk.* **59**, 1917.

(92) MARSHALL, *The Physiology of Reproduction,* 2nd ed. London, 1922.

(93) PEARL and PARSHLEY, *Biol. Bull.* **24**, 1913, p. 205.

(94) JOSS, *Arch. f. Tierheilk.* **43–44**, 1916–18.

(95) KAUPP, *Anatomy of the Domestic Fowl,* London, 1918.

(96) PALMER and ECKLES, *Missouri Agric. Exp. Stn. Res. Bull.* No. 10, 1914.

(97) ESCHER, *Zeitsch. Phys. Chem.* **83**, 1913, pp. 198–211.

(98) PALMER and KEMPSTER, *Journ. Biol. Chem.* **39**, 1919, p. 313.

(99) CORNER, Carnegie Institution of Washington, Pub. 222, *Contributions to Embryology,* II, No. 5, 1915.

(100) DELESTRE, *Journ. d'Anat. et de Physiol.* **46**, 1910, p. 286.

(101) ZSCHOKKE, *Schweizer Archiv f. Tierheilk.* **40**, 1898.

(102) SOBOTTA, *Anat. Hefte,* **8**, 1897.

(103) HAMMOND, *Journ. Agric. Science,* **11**, 1921.

(104) HEITZ, *Inaug. Diss.* Berne, 1906.

(105) KÄPPELI, *Inaug. Diss.* Berne, 1908.

(106) SANDES, *Proc. Linnean Soc. New South Wales,* **28**, 1903.

(107) CORNER, Carnegie Institute, Washington, Pub. 276, *Contributions to Embryology,* **13**, No. 64, 1921.

(108) BLAIR BELL, *The Principles of Gynaecology,* London, 1910, p. 87.

(109) SIMON, *Inaug. Diss.* Berne, 1904.

(110) ZIEGER, *Inaug. Diss.* Berne, 1908.

(111) GERLINGER, *Compt. rend. Soc. de. Biol.* **87**, 1922, p. 582.

(112) EMRYS-ROBERTS, *Roy. Soc. Proc.* B, **80**, 1908, p. 332.

(113) UTZ, *Tierärztliche Mitteil.,* Karlsruhe, 1890, quoted by Drahn, *Diss.* Hanover, 1913.

(114) KRUPSKI, *Schweizer Archiv f. Tierheilkunde,* **59**, 1917.

(115) SOMMER, *Zeitschr. f. Tiermedizin,* **16**, 1912.

(116) MAYER, *Zeitschr. f. Geburt. u. Gynaek.* **77**, 1915.

(117) SERVATIUS, *Inaug. Diss.* Berne, 1909.

(118) LOEB, *Amer. Journ. of Physiol.* **55, 56**, 1921.

(119) HILTY, *Schweizer Archiv f. Tierheilk.* **50**, 1908, pp. 268 and 353.

(120) ALBRECHTSEN, *The Sterility of Cows,* translation by Wehrbein, Chicago, 1917.

(121) ZUCHERKANDL, *Anat. Hefte,* **8**, 1897.

(122) PAIMANS, *Archiv f. Tierheilk.* **42**, 1916.

(123) BLAND SUTTON, *Surgical Diseases of the Ovaries and Fallopian Tubes*, London, 1891.

(124) BLAIR BELL, *The Principles of Gynaecology*, London, 1910.

(125) KELLER, *Anat. Hefte*, **39**, 1909.

(126) BAUMGÄRTNER, *Inaug. Diss.* Berne, 1910 and *Ost. Woch. f. Tierheilk. u. Rev. f. Tierheilk. u. Tierzucht*, 1911.

(127) FÜRSTENBERG, *Die Milchdrüsen der Kuh*, Leipzig, 1868.

(128) ZIETZSCHMANN, Chapter in Grimmer's *Chemie und Physiologie der Milch*, Berlin, 1910.

(129) RUBELI, *Schweizer Arch. f. Tierheilk.* **58**, 1916, p. 357.

(130) BITTING, *12th Ann. Report Indiana Agr. Exp. Stn.* Purdue Univ. 1900, p. 36.

(131) SCHIKELE, *Zeits. f. Morph. u. Anthrop.* **1**, 1899, p. 507.

(132) HENNEBERG, *Anat. Hefte*, Abt. 1, **25**, 1904, p. 683.

(133) BURCKHARD, *Anat. Hefte*, Abt. 1, **8**, 1897, p. 527.

(134) MACKENZIE and MARSHALL, *Journ. Agric. Sci.* **15**, 1925, p. 30.

(135) BELL, *Science*, N.S. **36**, No. 925, 1912.

(136) BEACH and CLARK, *Storrs (Conn.) Agr. Exp. Stn. 16th Ann. Report*, 1904, p. 131.

(137) PLUMB, *Judging Farm Animals*, London, 1916, p. 302.

(138) ECKLES, *Dairy Cattle and Milk Production*, New York, 1920, p. 25.

(139) ZWART, *Inaug. Diss.* Berne, 1911.

(140) WING, *Milk and its products*, New York, 1917, p. 3.

(141) WIRZ, *Arch. f. wiss. u. prakt. Tierheilk.* **39**, 1913, p. 375.

(142) PAULLI, *Laerebog i Huspattedyrenes Anatomi*, Copenhagen, 1919, p. 239.

(143) RIEDERER, *Arch. f. wiss. u. prakt. Tierheilk.* **29**, 1903, p. 593 and *Inaug. Diss.* Berne, 1903.

(144) NÜESCH, *Inaug. Diss.* Zürich, 1904.

(145) ZIETZSCHMANN, *Schweizer Arch. f. Tierheilk.* **59**, 1917, p. 645.

(146) HAMMOND, *Quart. Journ. Exp. Phys.* **6**, 1913, p. 311.

(147) GAVIN, *Quart. Journ. Exp. Phys.* **6**, 1913, p. 13.

(148) CHRIST, *Inaug. Diss.* Giessen, 1905.

(149) BITTING, *14th Ann. Report, Indiana Agr. Exp. Stn.* Purdue Univ. 1901.

(150) CROWTHER, *Abst. of Papers, Agric. Section of British Ass.*, Manchester, 1915.

(151) EMERY, *N. Carolina Agric. Exp. Stn. Bull.* **116**, 1895.

(152) STICKER, *Arch. f. mikros. Anat. u. Ent.* **54**, 1899, p. 1.

(153) HUG, *Inaug. Diss.* Zürich, 1906.

(154) MAŃKOWSKI, *Inaug. Diss.* Berne, 1903 and in *Polnischen Arch. f. biol. u. med. Wissenschf.* **2**, 1903, p. 7.

(155) KÄPPELI, *Inaug. Diss.* Zürich, 1918.

(156) ALDRICH and DANA, *Vermont Agr. Exp. Stn. Bull.* **202**, 1917.

(157) GRAVES, *Hoard's Dairyman*, No. 20, 1916.

(158) NELKE, *Inaug. Diss.* Berne, 1909.

(159) KEIM, *Inaug. Diss.* Leipzig, 1909 and *Ber. in d. Kgl. Tierärztl. Hochsch. z. Dresden*, N.S. 4, 1909, p. 286.

(160) SEITTER, *Inaug. Diss.* Zürich, 1910.

(161) DECHAMBRE, *La Vache Laitière*, Paris, 1912, p. 156.

(162) BAKKER, *Inaug. Diss.* Berne, 1909.

(163) ARMSBY, *The Nutrition of Farm Animals*, New York, 1917, p. 463.

(164) O'DONOGHUE, *Journ. of Physiol.* **43**, 1911–12 and *Quart. Journ. Microscop. Sci.* **57**, 1911.

(165) BOUIN and ANCEL, *Compt. rend. Soc. de Biol.* **67**, 1909 and *Journ. d. Phys. et de Path. gén.* **13**, 1911.

(166) SCHIL, *Recherches sur la Glande Mammaire*, Nancy, 1912.

(167) PEARL and SURFACE, *Maine Agric. Exp. Stn. Bull.* **237**, 1915.

(168) HILL, *Journ. Biol. Chem.* **33**, 1918.

(169) WOODMAN and HAMMOND, *Journ. Agric. Sci.* **12**, 1922, p. 97.

(170) HEAPE, *Proc. Phys. Soc., Journ. of Physiol.* **34**, 1906.

(171) LOWENTHAL, *Deut. Landw. Presse*, **34**, No. 72, 1907; and abs. in *Jahrb. wiss. u. prakt. Tierzucht*, 1908.

(172) OLIVER, *Edinburgh Medical Journal*, N.S. **9**, 1912, p. 530.

(173) LINDIG, *Zeits. f. Geburts. u. Gynaek.* **76**, 1914–15.

(174) HILL, *Journ. Dairy Science*, **2**, 1919.

(175) BARNOWSKY, *Deut. Landw. Tierzucht*, No. 6, 1912.

(176) LEBLANC, *Diseases of the Mammary Gland of Domestic Animals* (trans. by Nunn), London, 1904.

(177) SHEILD, *Diseases of the Breast*, London, 1898.

(178) FOGES, *Zentralbl. f. Physiol.* **19**, 1906, p. 233.

(179) STEINACH, *Pflüger's Arch. f. d. Physiol.* **144**, 1912, p. 71.

(180) ATHIAS, *Compt. rend. Soc. de Biol.* **79**, 1916, p. 553.

(181) MYERS, *Amer. Journ. Diseases of Children*, No. 5, 1919.

(182) JACKSON and STEWART, *Journ. Exp. Zool.* **30**, 1920.

(183) WOLZ, *Archiv f. Gynaek.* **97**, 1912.

(184) SCHOCHET, *Anat. Rec.* **10**, 1915–16, p. 447.

(185) ANCEL and BOUIN, *Compt. rend. Soc. de Biol.* **67**, 1909.

(186) KALTENEGGER, *Wiener Tierärztliche Monatsschr.* **2**, 1915, p. 12.

(187) VAN DER STRICHT, *Archives de Biol.* **27**, 1912, p. 585.

(188) VIGNES, *Physiologie Obstétricale*, Paris, 1923.

(189) MARSHALL and JOLLY, *Trans. Roy. Soc. Edinburgh*, **45**, 1907 and *Quart. Journ. Exp. Phys.* **1**, 1908.

(190) WINTZ, *Archiv f. Gynaek.* **113**, 1920.

(191) SONNENBERG, *Berliner Tierärztliche Wochenschrift*, No. 39, 1907, p. 700.

(192) SCHICKELE, *Bioch. Zeitschr.* **38**, 1912, p. 169.

(193) SOLORIEFF, *Rousski Vratch*, **11**, 1912, p. 466. Abs. in *Journ. de Phys. et de Path. gén.* **14**, 1912, p. 1077.

(194) WALLACE, *Farm Live Stock of Great Britain*, Edinburgh, 1923, p. 267.

(195) CRAMER, Chapter on "Biochemistry of the Sexual Organs" in Marshall's *Physiology of Reproduction*, 2nd ed. London, 1922.

(196) REGAUD and POLICARD, *Compt. rend. Soc. de Biol.* **53**, 1901, p. 449.

(197) VAN BEEK, *Diss.* Utrecht, 1921.

(198) LANE-CLAYPON, *Proc. Roy. Soc.* B, **77**, 1905.

(199) ASCHNER, *Archiv f. Gynaek.* **102**, 1914, p. 470.

(200) ATHIAS, *Archives de Biol.* **30**, 1919.

(201) POPOFF, *Archives de Biol.* **26**, 1911, p. 483.

(202) BARRINGTON, *Journ. of Anat. and Physiol.* **50**, 1915–16, p. 30.

(203) BÖHME, *Inaug. Diss.* Berne, 1909.

(204) MOREAUX, *Archives d'Anatomie Microscop.* **14**, 1913, p. 515.

(205) HILTY, *Schweizer Arch. f. Tierheilk.* **50**, 1908, pp. 268 and 353.

(206) SVEN WALL, *Meddels. f. Kgl. Vet. og Landbohøjskole*, Serum Lab., Copenhagen, **34**, 1914 and 10th Internat. Vet. Congress, London, 1914.

(207) KOLSTER, *Anat. Hefte*, Abt. 1, **20**, 1903, p. 233.

(208) HAMMOND, *Proc. Roy. Soc.* B, **89**, 1917, p. 534.

(209) WELLS, *Chemical Pathology*, London, 1918, p. 418.

(210) ANCEL and BOUIN, *Journ. d. Physiol. et de Path. gén.* **12**, 1910.

(211) KELLER, *Anat. Hefte*, **118**, 1909.

(212) TRAUTMANN, *Archiv f. Tierheilk.* **43–44**, 1916–18.

(213) HUTYRA and MAREK, *Spezielle Pathologie u. Therapie der Haustiere*, I, Jena,
1913, p. 762.

(214) RAEBIGER, *Journ. of Comparative Path. and Therap.* **20**, 1907.

(215) RETTERER, *Compt. rend. Soc. de Biol.* **44**, 1892, *Mémoires*, p. 101.

(216) LATASE, *Compt. rend. Soc. de Biol.* **45**, 1893, *Mémoires*, p. 135.

(217) MEYER, *Arch. f. Mikros. Anat.* **73**, 1909.

(218) RUDER, *Arch. f. path. Anat. u. Physiol.* (Virchow), **96**, 1884.

(219) DOHRN, *Arch. f. Gynaek.* **21**, 1883.

(220) KOCKS, *Arch. f. Gynaek.* **20**, 1882.

(221) ROEDER, *Arch. f. Tierheilk.* **24**, 1898.

(222) BARRINGTON, *Internat. Monatsschr. f. Anat. u. Physiol.* **30**, 1914, p. 1.

(223) MARSHALL, *The Physiology of Reproduction* (2nd ed.), London, 1922,
p. 268.

(224) KOCH, *Inaug. Diss.* Berne, 1909.

(225) PROFÉ, *Anat. Hefte*, **11**, 1899, p. 247.

(226) REIN, *Arch. f. Mikros. Anat.* **20**, 1881, p. 431 and **21**, 1882, p. 678.

(227) BONNET, *Anat. Hefte*, Abt. 2, **7**, 1897, p. 937.

(228) BRESSLAU, *The Mammary Apparatus of the Mammalia* (with a note by Hill),
London, 1920.

(229) LUSTIG, *Arch. f. Mikros. Anat.* **87**, 1917.

(230) ZSCHOKKE, *Arch. f. Mikros. Anat.* **93**, 1920.

(231) ERNST, MOHLER and EICHHORN, *Milk Hygiene*, Chicago, 1915, p. 10.

(232) MYERS, *Amer. Journ. of Anatomy*, **19**, 1916; **22**, 1917, and **25**, 1919.

(233) LOEB and HESSELBERG, *Journ. Exp. Med.* **25**, 1917.

(234) LENFERS, *Zeits. f. Fleisch- u. Milchhyg.* **17**, 1907, p. 340.

(235) MARTIN, Chapter in Ellenberger's *Handbuch der vergleichenden mikroskopischen
Anatomie der Haustiere*, I, Berlin, 1906.

(236) WILLIAMS, *Veterinary Obstetrics*, Ithaca (New York), 1917, p. 178.

(237) SAND, *Aarsskrift, Kgl. Vet. og Landbohøjskole*, Copenhagen, 1918.

(238) OPPERMANN, *Berlin. Tierärztl. Woch.* 1919, pp. 345 and 392.

(239) OPPERMANN, *Sterilität der Haustiere*, Hanover, 1922.

(240) REINHARDT, Series of diagrams published by M. and H. Schafer, Hanover.

(241) WILLIAMS, *The Veterinary Alumni Quarterly*, Ohio State University, 1914.

(242) WILSON, *The Veterinary Journal*, August, 1910, p. 460.

(243) SPENCER, *The Veterinarian*, **12**, 1839, p. 722.

(244) FRANCK-ALBRECHT, *Handbuch der Tierärztlichen Geburtshilfe*, Berlin, 1914, p. 154.

(245) WELLMANN, *Landw. Jahrb.* **39**, 1910, p. 409.

(246) EWART, *Trans. Roy. Soc. Edinburgh*, **51**, 1915.

(247) MARSHALL, *The Physiology of Reproduction*, 2nd ed. London, 1922, p. 50.

(248) SMITH, *Veterinary Physiology*, 5th ed. London, 1921, p. 783.

(249) BERRY HART, *Edinburgh Medical Journal*, Oct. 1913.

(250) FUHRIMANN, *Inaug. Diss.* Berne, 1906 and *Arch. f. Tierheilk.* **32**, 1906, p. 601.

(251) LIMMER, *Deutsche Landw. Tierzucht*, **17**, 1913 and *Inaug. Diss.* Dresden, 1912.

(252) STAPEL, *Inaug. Diss.* Berne, 1912.

(253) WEBER, *Deut. Tierärztl. Wchnschr.* Nos. 10, 11 and 12, 1910 and abs. in *Vet. Record*, No. 1139, 1910.

(254) RAVANO, *Arch. f. Gynaek.* **83**, 1907.

(255) MARCOTTY, *Arch. f. Gynaek.* **103**, 1914, p. 63.

(256) HAMMOND, *Proc. Roy. Soc.* B, **89**, 1917, p. 534.

(257) BERGMANN, *Arch. f. Tierheilk.* **47**, 1921, p. 292.

(258) ASSHETON, *Phil. Trans. Roy. Soc.* B, **198**, 1906.

(259) MEYER, *Arch. f. Gynaek.* **113**, 1920.

(260) KRAINZ, *Arch. f. Mikros. Anat.* **84**, 1914.

(261) BIENENFELD, *Biochem. Zeitschr.* **43**, 1912, p. 295.

(262) SMITH, *British Assoc.* 1910, p. 635; 1911, p. 414 and 1913, p. 670.

(263) EWART, *A critical period in the development of the horse*, London, 1897.

(264) MARSHALL and JOLLY, *Phil. Trans. Roy. Soc.* B, **98**, 1905.

(265) KLEINHAUS and SCHENK, *Zeitschr. f. Geburt. u. Gynaek.* **61**, 1907.

(266) CORNER, *Contributions to Embryology*, Carnegie Inst., Washington, **2**, No. 5, 1915.

(267) BLAIR BELL and HICK, *Brit. Med. Journ.* **1**, 1909, p. 655 and also BLAIR BELL, *The Sex Complex*, London, 1920, p. 30.

(268) FELLNER, *Arch. f. Gynaek.* **87**, 1909 and *Arch. Mikros. Anat.* **73**, 1909.

(269) LOEB, *Zentralbl. f. Physiol.* **25**, 1911.

(270) COLIN, *Traité de Physiologie Comparée des Animaux*, II, Paris, 1888, p. 988.

(271) ECKLES, *Univ. of Missouri Agri. Exp. Stn. Research Bull.* **35**, 1919.

(272) HAIGH, MOULTON and TROWBRIDGE, *Univ. of Missouri Agric. Exp. Stn. Res. Bull.* **38**, 1920.

(273) RÖRIK, *Arch. f. Tierheilk.* **33**, 1907, p. 421.

(274) BUCHEM, *Inaug. Diss.* Berne, 1909.

(275) KIRKHAM, *Anat. Rec.* **11**, 1916.

(276) MEYER, *Contributions to Embryology*, Carnegie Inst. Washington, II, No. 4, 1915.

(277) HAMMOND, *Journ. Agric. Sci.* **6**, 1914, p. 263.

(278) HAMMOND and APPLETON, *Growth and Development in the Sheep* (M.S.).

(279) HAMMOND, *Journ. Agric. Sci.* **11**, 1921, p. 337.

(280) HART, STEENBOCK and HUMPHREY, *Wisconsin Agri. Exp. Stn. Res. Bull.* **17**, 1911 and **49**, 1920.

(281) ECKLES, *Univ. of Missouri Agri. Exp. Stn. Res. Bull.* **26**, 1916.

(282) ANCEL and BOUIN, *Compt. rend. Soc. de Biol.* **72**, 1912.

(283) BOYD, *Journ. of Heredity*, **5**, 1914 and *Proc. Amer. Breeders Ass.* 1908.

(284) Crew, *Proc. Roy. Soc.* B, **95**, 1923, p. 228.

(285) Williams, *New York State (Cornell Univ.) Veterinary Coll. Report*, 1915–16, p. 117.

(286) Williams, *U.S. Dept. of Agric. Bull.* No. 106, 1914.

(287) Rab, *Inaug. Diss.* Utrecht, 1903.

(288) Ancel and Bouin, *C. r. Ass. des Anat.* 13ᵉ réun. Paris, 1911 and *Compt. rend. Soc. de Biol.* **73**, 1912.

(289) Fraenkel, *Arch. f. Gynaek.* **99**, 1913.

(290) Bruyn-Ouboter, *Inaug. Diss.* Berne, 1911.

(291) Gamgee, *Brit. and For. Med. Chir. Review*, 1864.

(292) Marshall, *The Physiology of Reproduction*, 2nd ed. London, 1922, p. 432.

(293) Kolster, *Anat. Hefte*, Abt. 1, **18**, 1902, p. 457.

(294) Schauder, *Arch. f. Tierheilk.* **46**, 1921, p. 187.

(295) Jenkinson, *Proc. Zool. Soc. London*, 1906, **1**, p. 73.

(296) Oliver, *New York Medical Journal and Medical Record*, July, 1922.

(297) Ledermann, *Inaug. Diss.* Berlin, 1903.

(298) Pomayer, *Das Zurückhalten der Nachgeburt beim Rind*, Berlin, 1919.

(299) Fraenkel, *Arch. f. Gynaek.* **55**, 1898, p. 269.

(300) Schäfer, Quain's *Anatomy*, **2**, Pt. 1, *Textbook of Microscopic Anatomy*, London, 1912, p. 107.

(301) Heape, *Roy. Soc. Phil. Trans.* B, **185**, 1894.

(302) Schmaus, *Textbook of Pathology and Pathological Anatomy*, trans. by Tharyer and Ewing, London, 1903, p. 86.

(303) Jenkinson, *Vertebrate Embryology*, Oxford, 1913.

(304) Romolotti, *Exp. studies on cattle and pigs at Royal Zootech. Inst. Reggio Emilia*, 1911 (abs. in *Int. Inst. of Agric. Bull. of Agric. Intell.* 1912, p. 471).

(305) Lehmann, *Arch. f. Tierheilk.* **48**, 1922, p. 233.

(306) Fellner, *Arch. f. Gynaek.* **100**, 1913, p. 641.

(307) Schröder and Goerbig, *Zeitschr. f. Geburts. u. Gynaek.* **83**, 1920–21, p. 764.

(308) Schultze, *Inaug. Diss.* Berne, 1909.

(309) Schwarz, *Inaug. Diss.* Dresden, 1912.

(310) Müller, *Inaug. Diss.* Zürich, 1919.

(311) Lillie, *Journ. Exp. Zool.* **23**, 1917, p. 371.

(312) Lüer, *Deut. Landw. Tierzucht*, **17**, 1913, p. 254.

(313) Keller, *Jahrb. f. Tierzucht*, **10**, 1916.

(314) Stålfors, *Monat. f. prakt. Tierheilk.* **27**, 1916, p. 338.

(315) Leopold, *Arch. f. Gynaek.* **15**, 1880, p. 258.

(316) Küpfer, *Vierteljahrssch. d. Naturfor. Gesell. in Zürich*, **65**, 1920, p. 377.

(317) Schauder, *Arch. f. Anat. u. Physiol. Anat.* Abt. 1912, p. 259.

(318) Crew and Fell, *Q.J. Microscop. Sci.* **66**, 1922, p. 557.

(319) Nauta, *Inaug. Diss.* Berne, 1906.

(320) Bernard, *Ann. Sci. Nat.* **10**, 1858 and *Leçons sur les phénomènes de la vie*, Paris, 1879.

(321) Dennhardt, *Deutsche Tierärzt. Wochensch.* 1906, p. 308.

(322) Retterer and Lelièvre, *Journ. d'Anat. et de Physiol.* **50**, 1914–19, p. 342.

(323) Woodman and Hammond, *Journ. Agric. Sci.* **13**, 1923, p. 180.

(324) Kopf, *Inaug. Diss.* Berne, 1909.

(325) Bernhard-Vonnahme, *Inaug. Diss.* Hanover, 1912.

(326) SASSENHAGEN, *Inaug. Diss.* Berne, 1910.

(327) WOODMAN, *Biochem. Journ.* **15**, 1921, p. 187.

(328) GAVIN, *Journ. of Agric. Sci.* **5**, 1913, p. 309.

(329) HAMMOND and SANDERS, *Journ. Agri. Sci.* **13**, 1923, p. 74.

(330) BROUHA, *Arch. de Biol.* **21**, 1905, p. 459.

(331) BENOIT, *Compt. rend. Assoc. des Anatomistes*, 27ᵉ réun. April, 1922.

(332) BOUIN and ANCEL, *Journ. de Phys. et de Pathol. générale*, **13**, 1911.

(333) MACKENZIE, MARSHALL and HAMMOND, *Journ. Agric. Sci.* **6**, 1914, p. 182.

(334) LOEB and HESSELBERG, *Journ. Exp. Med.* **25**, 1917.

(335) BOUIN and ANCEL, *Compt. rend. Soc. de Biol.* **72**, 1912, p. 129 and **76**, 1914, p. 150.

(336) ASCHNER and GRIGORIU, *Arch. f. Gynaek.* **94**, 1911, p. 766.

(337) FELLNER, *Arch. f. Gynaek.* **100**, 1913, p. 641.

(338) LANE-CLAYPON and STARLING, *Proc. Roy. Soc.* B, **77**, 1906.

(339) BIEDL and KÖNIGSTEIN, *Zeits. f. Exp. Path. u. Therap.* **8**, 1911, p. 358.

(340) FRANK, *Surg. Gynaec. and Obstetrics*, **25**, 1917, p. 329, quoted from *Exp. Stn. Rec.* **41**, p. 173.

(341) O'DONOGHUE, *Quart. Journ. Microscop. Sci.* **57**, 1911–12, p. 187.

(342) MARSHALL and CROSLAND, *Journ. of the Board of Agric.* **24**, 1918, p. 1358.

(343) MACOMBER, *Boston Medical and Surgical Journal*, **187**, 1922, p. 297.

(344) LEWIS, *Oklahoma Agric. Exp. Stn. Bull.* **96**, 1911, p. 21.

(345) SAKOWSKY, quoted by Huish, *Sterility in Mares, Cows, and Bitches*, London, 1910, p. 91.

(346) STRODTHOFF, *Arch. f. Tierheilk.* **48**, 1922, p. 28.

(347) BEAVER, BOYD and FITCH, *Journ. Amer. Vet. Med. Ass.* Aug. 1922, p. 469 (abs. in *Vet. Record*, Oct. 14th and 28th, 1922).

(348) KELLER, *Wiener Tierärztl. Monatsschr.* **1**, 1914, p. 11.

(349) FITCH, *New York State Veterinary Col. Report*, 1915–16, p. 199.

(350) ZSCHOKKE, *Die Unfruchtbarkeit des Rindes*, Zürich, 1900; see also (40) and *Schweizer Arch. f. Tierheilk.* **40**, 1898, p. 253.

(351) BONGARDT, *Arch. f. Tierheilk.* **47**, 1921, p. 15.

(352) SCHUMANN and HIERONYMI, *Arch. f. Tierheilk.* **40**, 1914, p. 193.

(353) FRAENKEL (L.), *Arch. f. Gynaek.* **56**, 1898, p. 355.

(354) FRAENKEL (E.), *Arch. f. Gynaek.* **57**, 1899, p. 511.

(355) PUGH, *Veterinary Journal*, **78**, 1922.

(356) LOTHE, *North American Veterinarian*, **3**, 1922.

(357) HINRICHS, *Inaug. Diss.* Hanover, 1919.

(358) LAURIE, *Dept. of Agric. South Australia, Bull.* No. 72, 1912.

(359) MEYER, *Arch. f. Gynaek.* **100**, 1913.

(360) COHN, *Arch. f. Gynaek.* **87**, 1909.

(361) BURGHARDT, *Arch. f. Tierheilk.* **37**, 1911, p. 455.

(362) STAZZI, *Veterinary Journal*, No. 440, 1912.

(363) PETERS, *Nebraska Agric. Exp. Stn. Report*, 1909.

(364) HOBDAY, *Castration and Ovariotomy*, 2nd ed. Edinburgh, 1914.

(365) REISINGER, *Wiener Tierärztl. Monatsschr.* **1**, 1914, p. 16.

(366) SCHEIDEGGER, *Die Sterilität des Rindes*, Berne, 1914.

(367) POSSELT, *Inaug. Diss.* Hanover, 1914.

(368) SKELLETT, *Parturition in the Cow*, London, 1807.

(369) WILLIAMS, *The diseases of the Genital Organs of Domestic Animals*, Ithaca (New York), 1921.

(370) MCNUTT, *Journ. Amer. Vet. Med. Ass.* **65**, No. 5, 1924.

(371) MURPHEY, *Journ. Amer. Vet. Med. Ass.* **65**, No. 5, 1924.

(372) MURPHEY, MCNUTT, ZUPP and AITKEN, *Journ. Amer. Vet. Med. Ass.* **67**, No. 3, 1925.

(373) MURPHEY, MCNUTT, ZUPP and AITKEN, *Journ. Amer. Vet. Med. Ass.* **68**, No. 4, 1926.

(374) MURPHEY, MCNUTT, ZUPP and AITKEN, *Veterinary Medicine*, Aug. 1925.

(375) MURPHEY, *Veterinary Medicine*, June, 1925.

(376) STÄHELI, *Festschrift z. 70 Geburts. v. Prof. E. Zschokke*, Zürich, 1925.

(377) WILLIAMS and WILLIAMS, *N. Amer. Vet.* **4**, 1923, pp. 232 and 305.

(378) HAIGH, MOULTON and TROWBRIDGE, *Missouri Agr. Exp. Sta., Res. Bul.* **38**, 1920.

(379) FINCHER and WILLIAMS, *Cornell Vet.* Jan. 1926.

(380) GOWEN, *Biol. Bul.* **41**, 1922.

(381) COLE and RODOLFO, *Amer. Soc. for Animal Prod., Proc. Ann. Meeting*, 1924.

(382) BASCOM, *Amer. Journ. Anat.* **31**, 1923, p. 223.

(383) WOODMAN and HAMMOND, *Journ. Agric. Sci.* **15**, 1925, p. 107.

(384) FREI, *Spez. Path. Anat. der Haustiere, IV Milchdrüse, Weibliche Geschlechtsorgane*, Berlin, 1925.

(385) ALLEN and DOISY *et al.*, *Amer. Journ. Physiol.* **69**, 1924; *Amer. Journ. Anat.* **34**, 1924; and *Journ. Biol. Chem.* **61**, 1924.

(386) COURRIER, *Compt. rend. Soc. de Biol.* **90**, 1924 and **93**, 1925.

(387) LOEB, *Amer. Journ. Anat.* **32**, 1923, p. 305.

(388) CORNER, *Contributions to Embryology*, No. 75.

(389) WATRIN, *Thesis*, Liège, 1924.

(390) SEITZ and WINTZ, *Münch. Med. Woch.* 1919, p. 475.

(391) PARKES, *Proc. Roy. Soc.* B. **100**, 1926, p. 172.

(392) KAMM, *Diss.* Bern, 1925.

(393) BRUN, *Diss.* Bern, 1921.

(394) DONNET, *Diss.* Bern, 1922.

(395) ASDELL, *Journ. Agric. Sci.* **15**, 1925, p. 358.

(396) DRUMMOND-ROBINSON and ASDELL, *Journ. Physiol.* **61**, 1926.

(397) WESTER, *Eierstock und Ei*, Berlin, 1921.

(398) KNAUS, *Journ. Physiol.* **61**, 1926, p. 383.

(399) DIXON and MARSHALL, *Journ. Physiol.* **59**, 1924, p. 276.

(400) WARWICK, *Anat. Rec.* **33**, 1926, p. 29.

ADDENDUM

Skellett (368) in a book published in 1807 noted in the cow the cervical seal and a mucous layer lining the amnion which became fluid about the 6th–7th month of pregnancy. Williams (369) has given a good description of many of the changes occurring in the reproductive organs of the cow in their relation to pathology. While his explanations in the main agree with those given above we believe that some which are described as pathological may be treated rather as physiological, for example the cystic stage of the corpus luteum, the occurrence of lymphatic nodules in the vagina, and the apical necrosis of the foetal sac with surrounding endometritis; the latter is probably due to the non-vascularisation of the ends and the accumulation of deposits of uterine milk which is not absorbed in this area.

In a series of papers Murphey, McNutt, Zupp and Aitkin (370–375) have described many of the changes occurring during the reproductive cycle in the cow which are in the main in agreement with those described above; their chart showing the size changes in the follicle and corpus luteum during the ovarian cycle corresponds very closely with our own. They also give many illustrations of the histological structure of the reproductive organs in the cow; the time of ovulation is given as 30–65 hours after the onset of heat and they state that they have used with success injections of oestral hormone for the cure of sterility. Stäheli (376) also claims to have obtained successful results in the treatment of sterility in praesenile cows by the transplantation of ovaries, while Williams and Williams (377) state that good results have been obtained by treating nymphomaniac cows (ovarian cysts) with injections of adrenal and pituitary extract; they state that the condition is most frequently found in high milking cows which are well fed. Haigh, Moulton and Trowbridge (378) mention that heifers subject to a low plane of nutrition failed to come on heat promptly. Arrested development of the Müllerian ducts in cattle associated with inbreeding has been described by Fincher and Williams (379). Data on the occurrence of identical twins in cattle have been published by Gowen (380), while Cole and Rodolfo (381) find the relative frequency of twins is greatest in cows calving from July to September. Bascom (382) could find no interstitial cells in the cow's ovary until the follicles began to undergo atrisia. A chemical investigation of the mucin of the cow's cervix in the cycle and during pregnancy has been published by Woodman and Hammond [383]. A description of the reproductive organs and udder of the cow in relation to pathology has been published by Frei [384]; he states that the maximum hypertrophy of the uterine mucosa occurs at the 12th day after the beginning of heat and believes that an hypertrophy of the uterine mucosa is associated with the persistence of the corpus luteum; illustrations of the amniotic pustules are given and he states that not only do 75 per cent. of heifers and 50 per cent. of cows bleed in the uterus at heat but that bleeding may also occur in the uterus during the last week of pregnancy.

As to the cause of heat Allen, Doisy *et al.* [385] state that they have separated an extract from the follicle which when injected will produce the symptoms of heat and that the hormone is not species specific. Courrier [386] and Dr S. A. Asdell have also obtained similar results. Loeb [387] is of the opinion that while the internal secretions of the follicle act on the vagina as well as to a limited extent on the uterus those of the corpus luteum act on the uterus and mammary gland but not on the vagina. Corner [388] found in monkeys that while ovulation usually occurs in a definite relation to menstruation, *i.e.* 12–14 days before the next menstrual flow, the latter can also occur at times without ovulation. Watrin [389] from a study of the cycle in woman finds that removal of the corpus luteum causes menstruation 24–48 hours afterwards, while Seitz and Wintz [390] have shown that the application of X-rays which destroys the follicle but not the corpus luteum results in the absence of the next menstrual flow if it is applied shortly after menstruation, *i.e.* during the follicular phase, but if it is applied during the end of the cycle, *i.e.* during the luteal phase one more menstrual flow occurs. This suggests that the menstrual flow is a pseudo-pregnant rather than a pro-oestral effect. Parkes [391] however finds that in mice the symptoms of heat may occur regularly after the follicular system has been destroyed by X-rays.

The histological structure of the cow's teat has been described by Kamm [392] while biometrical studies of the escutcheon have been made by Brun [393] and Donnet [394]. Asdell [395] has described the changes in amount and composition of the fluids from the udders of heifers and goats during the "upcurve" of lactation in the first pregnancy; he considers that the globulin is produced when the gland cell changes from the growth to the secretory phase of its life. Drummond-Robinson and Asdell [396] find that if, in goats pregnant for the first time, the corpus luteum is removed (with abortion as a result) before the thick secretion appears, no milk is produced, while if it is removed after the thick secretion appears milk is produced. Wester (397) states that in pregnant cows removal of the ovary containing the corpus luteum leaving the other intact results in death of the foetus.

The growth changes in the pregnant uterus are referred to by Knaus (398) who found that the effect of injections of pituitary extract into pregnant rabbits varied according to the stage of pregnancy at which they were made; no effect was produced by injections up to about the 18th day of pregnancy, while between this time and about the 28th day the foetuses were killed and abortion occurred a few days later, but injections made from this time to the end (32nd day) of pregnancy resulted in immediate birth of the litter. The pituitary secretion as a probable factor in parturition had previously been suggested by Dixon and Marshall (399). Experimental evidence of the occurrence of intra-uterine migration of the ovum in pigs has been published by Warwick (400).

INDEX OF AUTHORS

INDEX OF SUBJECTS

Ovum, degeneration of, 129; migration of, 160

Parturition, oestrus after, 6, 15, 123; uterus after, 53; cervix during, 55, 165; symptoms of, 119 et seq.; corpus luteum after, 123; factors in, 136; foetal membranes after, 153, 154; presentation at, 164; cervix after, 166; and pituitary, 217
Pelvic ligaments, see *Sacro-sciatic ligaments*
Penis, in coitus, 54
Persistent corpora lutea, removal of, 16, 17; and rhythm, 16; causes of, 126; and sterility, 184, 185
Pigment, in uterus, 51; in cotyledons, 88; in pregnancy, 149, 150
Placenta, causes of formation, 154, 155
Pregnancy, and rhythm, 16; comparison with cycle, 90, 120; description of, 112 et seq.; diagnosis of, 113 et seq.; duration of, 118 et seq.; ovaries, 121 et seq.; ovary weights, 125; follicular atrophy, 128; uterus, 130 et seq., 136, 138 et seq.; foetal fluids, 134, 135; uterine glands, 141, 142, 143; cotyledons, 147 et seq.; cervix, 164 et seq.; vagina, 166; vaginal secretions, 167; mammary glands, 168 et seq.
Pro-oestrum, absence of, 29
Pseudo-pregnancy, and pregnancy, 120
Puberty, age of, 7, 8; follicle, 42; ovary, 43, 45; uterus, 52; uterine glands, 92; mammary ducts, 172

Rabbit, age of puberty, 8; ovulation, 14; absence of rhythm, 16, 17, 20; hysterectomy, 24; removal of corpus luteum, 28, 127; corpus luteum, 40; corpus luteum in pregnancy, 41, 122; uterus, 59, 140; mammary gland, 68, 110, 176, 177, 178; bleeding into follicle, 77; injection of extracts, 81, 155; interstitial cells, 84; deciduomata, 89, 154; uterine glands, 90; pseudo-pregnancy, 120; persistence of corpus luteum, 126; follicular atrophy, 128, 129; foetal growth, 133; myometrial gland, 134, 139; foetal absorption, 137, 138; attachment, 154; migration of ovum, 160; cervix, 165; pituitary on pregnancy, 217
Rat, cycle, 14; corpus luteum extracts, 16; corpus luteum, 27, 28, 40, 126; uterus, 50; mammary gland, 72, 110; ovulation, 78; interstitial cells, 84, 85; vaginal cycle, 97 99
Ratios, ovarian weights in pregnancy, 125; parts of pregnant uterus, 130 et seq.
Red Polls, nipple size, 64
Rhythm, of oestrus, 16

Sacro-sciatic ligaments, at parturition, 120; and nymphomania, 120, 188
Service, and cycle, 14; and oestrus, 20;

contractions after, 30; straining after, 31; and bleeding, 59, 96
Service periods, and time of year, 5; and nutrition, 7; and age, 7
Sex ratio, and oestrus, 35
Shorthorns, in experiments, 2; corpus luteum, 45; position of uterus, 114; foetal growth, 158
Simmentals, uterus weight, 51; duration of pregnancy, 118; atrophic follicles, 129
Sow, corpus luteum, 16, 41, 80, 123, 125; cycle and oestrus, 21; follicle, 45; ovulation, 78; cervix, 94, 165; foetal weight, 133; attachment, 145, 146; migration of ovum, 160, 217; mammary pigment, 176
Spermatozoa, in vasectomy, 4; and cervix, 94; life of, 119
Steppe cattle, age of puberty, 8; atrophic follicles, 129
Sterility, and cystic corpora lutea, 39, 40; account of, 179 et seq.; diagnosis of, 193, 194; frequency of, 194
Stigma, of follicle, 35
Stratum vasculosum, of uterus, 140
Stricht-canal, muscle in, 63, 64; formation of, 105, 108
Suckling, oestrus during, 15
Supernumerary nipples, position of, 59; frequency of, 60; in bulls, 60

Teats, see *Nipple*
Testes, in vasectomy, 4
Theca externa, structure, 76
Theca interna, structure, 76, 77; after ovulation, 79, 83
Theca luteal cells, structure, 76, 77; in corpus luteum, 83; in stroma, 84
Time of year, conceptions during, 5; of calving, 5; and puberty, 8; and cycle, 11; and oestrus, 19; and duration of pregnancy, 118
Trophoblast, in crypts, 141, 149; functions, 151, 152, 156
Tuberculosis, and cycle, 11; and follicle, 17
Tunica albuginea, and ovulation, 35
Twin pregnancies, frequency, 41, 215; corpora lutea in, 41, 127; foetal fluids in, 136, 137

Udder, see *Mammary gland*
Under-nutrition, and cycle, 12, 17; and ovaries, 82; and oestrus, 215
Uterine artery, in pregnancy, 116
Uterine bleeding, causes, 28
Uterine epithelium, in pregnancy, 151, 152, 156
Uterine glands, at oestrus, 50; mucin in, 86; distribution, 87; in the cycle, 89, 90, 91; involution of, 91, 143; before puberty, 92; in pregnancy, 140 et seq.
Uterine milk, in pregnancy, 140 et seq.; and cotyledons, 149, 151

Printed in the United States
By Bookmasters